JAMES ROBERT BROWN

WHO RULES IN SCIENCE?

AN OPINIONATED GUIDE TO THE WARS

HARVARD UNIVERSITY PRESS
Cambridge, Massachusetts
London, England
2001

Printed in the United States of America

Library of Congress Cataloging-in-Publication Data

Brown, James Robert.
Who rules in science? : an opinionated guide to the wars / James Robert Brown.
p. cm.
Includes bibliographical references and index.
ISBN 0-674-00652-6 (alk. paper)
1. Science–Social aspects. 2. Science and state. I. Title.

Q175.5 .B758 2001
306.4'5–dc21

2001026461

To the memory of my mother
Isabel Brown (née Denovan)
1918–1991

Contents

Preface

There is something quite old about the so-called science wars. People have been fighting over the nature of knowledge since the beginning of intellectual life. Protagoras said that man is the measure of all things. Plato was appalled and spent much of his life combating relativism and other early forms of social constructivism. After the stunning successes of the Scientific Revolution, the Enlightenment embraced the twin ideals of progress and rationality. The sentiment was celebrated in Pope's famous couplet:

> Nature, and Nature's Laws lay hid in Night.
> God said, *Let Newton Be!,* and All was *Light.*

But the romantic rebellion followed on quickly. When Keats complained that "Philosophy would clip an angel's wings," he was rejecting any scientific understanding of the world that would rob it of its enchanting mystery.

There is also something quite new about the science wars. The mix of science, epistemology, and politics is exceedingly novel, exceedingly interesting, and exceedingly important.

How novel? A mix of epistemology and politics in science is hardly unprecedented; rather, it's the different ways that a divided Left looks at the issues that is strikingly new.

How interesting? If, halfway through the book, you're not completely caught up in these issues, then (presuming you're not a complete dullard) you will have only me to blame for making a botch of it.

How important? The way we live hangs on it. This many-sided debate is about *who rules* and *who should rule*—not just in science, but generally.

It wasn't so long ago that C. P. Snow wrote his famous *Two Cultures and the Scientific Revolution,* a work often cited in the current round of the science wars. But what gets cited is Snow's lament that the two cultures–scientific and literary–don't much understand one another. This is only a small part of Snow's message. His other claims are much more important. He declared, for instance, that the literati are instinctively on the political Right and have been responsible for some of the great horrors of our times. Scientists, he claimed, are instinctively on the Left and would do a much better job of solving social problems. In short, Snow's little book is about who should rule.

Today's version of the science wars exactly reverses things. One famous polemic refers to the critics of science as "the academic left" and the Right is identified–if only implicitly–as pro-science. Aside from this striking reversal of positions, much of the current debate is still about political power.

By phrasing it "Who should rule?" am I not just offering a choice of dictators? I don't mean to do this, and the various participants in the science wars (I presume) don't mean to do this either. Perhaps "Whose advice should we heed?" would be a more democratically acceptable way to put it. I prefer the punchier expression.

The dichotomy of an anti-science Left against a pro-science Right is a common perception. Snow misread his scientists (in 1959) and we very likely misread ours today. The real value of the now infamous Sokal affair is to bust this simple-minded dichotomy and give some elbow room to a left-wing alternative that is (with important qualifications) broadly pro-science. One thing that Sokal did not do and did not try to do is to separate science from political concerns. He is as much concerned with who should rule as anyone is. The battleground in the current round of the science wars is epistemology (What is evidence? Objectivity? Rationality? Could any belief be justified?), and that is the focus of most of this book. The stakes are political, however; social issues are constantly lurking in the background. How we structure and organize our society is the consequence. Whoever wins the science wars will have an unprecedented influence on how we are governed.

My title is *Who Rules in Science?* I will abbreviate this to *Who Rules?*, a locution drawn from current adolescent jargon. It means "Who is best?" But I like the ambiguity–it can also mean "Who does rule?" in the political sense. It can even suggest the question, "Who should rule?" This book is an introduction–an opinionated introduction–to the issues, written for the general reader. There are a few places where the going is slow, but I have tried to make it both readable and informative. I'd also like to convince readers of the underlying theme, which I could simply, if naively, express thus: Great science and social justice are linked–intelligent high-mindedness anywhere spills over everywhere.

There are many different battles that could be called "science wars." I'm not talking about them all. In particular, there are ongoing battles between science and the religious Right. There is an uneasy relationship between science and parts of the environmental movement. There is an alarming increase in the commercialization of science; the biotechnology industry is funding research and patenting knowledge everywhere to the possible detriment of society in general. These vitally important topics are discussed briefly in the Afterword. They are not my principal focus, however, even though they often overlap with my concerns about epistemology and its link to politics. What is the connection between my topic and the others? To understand the science wars in their entirety, we first need to explore the issues of objectivity, values, and social influences discussed in this book. Then we can move on.

This volume is dedicated to my mother. It's only with hindsight that I see how much she influenced me, especially in ways that have a bearing on the subject at hand: she taught me to love science (and intellectual life in general) and to be a feminist. For both of these I am eternally grateful. Trying to see things from other perspectives, however alien and unnatural they may seem, is the most valuable thing I learned from her. Not only is it morally just, but it has a great epistemic advantage, too, since one might actually learn something new. Though she didn't conceptualize it this way at the time, she taught me to see a connection between politics and epistemology, a connection completely compatible with genuine progress on both fronts.

Acknowledgments

I wish to thank Ric Arthur (and his students), Robert Barsky, Lars Bergström, Kate Brick (for exceptionally thorough copy-editing), Martin Curd, Michael Fisher (a wonderfully helpful editor who made everything easy), Karyn Freedman, Bill Frucht, Yves Gingras, Ian Hacking, Mary Leng, Josh Mozersky, Kathleen Okruhlik, Michael Ruse, Alan Sokal, Mary Tiles, students in my philosophy of science classes in which I tried out some of this material, the anonymous readers and the syndics of Harvard University Press, and the Social Sciences and Humanities Research Council of Canada (SSHRC) for its continuing financial support. I must add that Ric Arthur, Martin Curd, Robert Barsky, Lars Bergström, Michael Ruse, and Alan Sokal each provided me with extremely detailed and valuable criticisms and comments on almost every page. Even if I did not always take their advice, I hope my sincere gratitude matches my very great debt to each of them. Finally, I must thank Bijon Roy for proofreading and for making the index.

WHO RULES IN SCIENCE?

1
Scenes from the Science Wars

A Trip to the Museum

In Canada the science wars are fought by conscientious objectors. I recently went to the Ontario Science Centre in Toronto to see a special exhibit, "A Question of Truth." The aim was to make us aware of the impact of different "points of view" on theorizing and to raise questions about social and cultural influences on science. I was especially drawn to it by an article in *The Globe & Mail* by Stephen Strauss, the science editor. He complained that the exhibit sent out a "relativistic" message and was, he suspected, motivated by "political correctness."

What surprised me was just how conservative most of the exhibit was. Strauss couldn't have been further off the mark–at least as far as relativism goes. The whole thing was polite, cautiously liberal, and designed to give no offense to anyone, including scientists. One of the displays (there are forty in all), titled "East Meets West," concerns acupuncture. A large plastic body illustrates the procedure; an accompanying text says that western science has made discoveries about nerve endings close to the skin and how stimulating them triggers the release of endorphins in the brain. It goes on to say that this may explain why acupuncture therapy has helped chronic pain sufferers. Notably, there is no mention of eastern explanations of how acupuncture works.

Another exhibit, "Secrets of the Heart," asks "Who revealed them first?" The display notes that an Islamic physician, Ibn al-Nafis of Syria, described the process of how oxygen enters the blood in the thirteenth century–300 years before anyone in Europe. The moral is that not all great scientific accomplishments are European.

Both of these exhibits reflect a highly conservative and quite traditional view of science–one completely compatible with the most staunchly anti-social constructivist versions of scientific realism. In each case the view is this: there is a way things are–about pain and endorphins, about blood and oxygen–and current, Western science correctly describes them. There is not a hint of relativism about this. "Political correctness"? Yes, it's there–in the form of giving due credit to those non-Europeans who have discovered the truth. Most so-called political correctness is innocuous in just this way; it's hard to understand why anyone would oppose it. And if this is what social constructivism is, it's hard to see why anyone would oppose it either.

The situation is ironic. Political correctness often irritates people because it sometimes appears to embrace a strong form of social constructivism. Yet someone who is a serious constructivist would very likely *deny* the alleged prior discovery of how oxygen gets into the blood. After all, what is "oxygen"? One need only read Thomas Kuhn's very influential *Structure of Scientific Revolutions* on this very term. He discusses the history of chemistry–Lavoisier vs. Priestley–and claims that it is highly problematic to take a concept like *oxygen* and apply it in distinct theoretical contexts. The museum's version of political correctness is so conservative, in fact, that it rides rough-shod over Kuhnian incommensurability, a primary source of relativistic inspiration. Looking at an exhibit like this, one might even say with Zola, commenting on the Dreyfus affair: *La vérité est en marche; rien ne peut plus l'arrêter.*

Some exhibits at the Science Centre have a distinctly political tone to them. One of these is on "Race and Prejudice." It includes a very good short film on race and IQ studies, eugenics, and related issues. It does not simply declare that certain lines of research were racist. Rather, it wisely points out that particular lines of research are in serious conflict with other things established by regular, orthodox science. In short, *good science fights socially pernicious science*–a theme I always find heart-warming.

Perhaps the most interesting and provocative exhibit is the display "Sex and Science," which asks: "Does your sex determine your point of view? How might this affect scientific research?" There's a picture of a human egg surrounded by sperm cells. Visitors are asked to choose between statements *a* and *b* as best representing their beliefs. Although there is a long list of alternative pairs, the first pair is representative:

a. This is a human egg waiting to be fertilized by fiercely competitive sperm cells, each one a formidable .06 mm weapon, tipped with a chemical warhead.
b. This is a wastefully huge swarm of sperm flailing aimlessly around a mucus-enshrouded egg. Eventually the egg yanks one reluctant sperm inside so it can fertilize itself and become an embryo.

Of course, these are provocatively posed alternatives, and we are likely to reject both. Nevertheless, when no other alternatives are offered, the results can be divisive. And the statistics make it plain. A record is kept of the answers and of the gender of the person responding. Males tend to pick *a*, females tend to pick *b*. The "Sex and Science" display dramatically makes the point so many feminist critics have stressed: gender influences research. But a distinction quickly leaps to mind: gender influences lines of research and perhaps what people *tentatively believe* to be true, but it does not influence the truth itself. At most one of the *a-b* pair is true, perhaps neither–but certainly not both. Holding both true is a kind of relativism we often hear about, but few (if any) feminist writers on science actually maintain such a view.

If there is a moral that comes out of the exhibit, "A Question of Truth," it could only be that *point of view* matters to the direction of research, but not to the facts themselves. This is a rather liberal view of science, but quite within the realist camp and quite in harmony with scientific orthodoxy. It is also, I suspect, a reasonable reflection of the educated public's view of science, as well as a view with which most working scientists can happily live.

As I mentioned already, Canada's version of the science wars is waged by pacifists. For real bloodletting, we need to look elsewhere. Before we turn to C. P. Snow's famous book, *Two Cultures,* I will slip in a word about key terms. There is no uncontroversial definition of "social construction," "relativism," or other important terms that are scattered throughout this book, not even "Left" and "Right." But rough definitions might be useful. To say that knowledge is a *social construction* is to say that it is the product of various social factors and not the result of an objective investigation into how things are independent of our social interests. But there is more to it than just belief–there are no objective facts of the matter to be discovered, according to constructivists.

It needn't be an all or nothing matter. Some social constructivists apply the view to everything, but one could–with considerable plausibility–claim that quantum mechanics, for example, is objective knowledge, whereas economics is largely a social construction.

Relativism is related to social construction and often taken to be a consequence of it. It says that knowledge (scientific, moral) is tied to a group or society ("Polygamy is morally proper for them, but wrong for us." "The Big Bang is factually true for us, but for them the world started in a different way, and they're correct, too."). There is no moral right or wrong, no factual truth or falsehood over and above what is accepted by a particular society. The old saw: "When in Rome, . . ." doesn't do relativism complete justice, but it captures some of its flavor.

The *Left-Right* dichotomy stems from the French Revolution, when the more progressive members of the National Assembly sat *à gauche* while their opponents were seated *à droit.* The terms have stuck and are used in a common, though loose, fashion throughout the world. Left-wingers today would often include opposition to racism, opposition to sexism, pro-environmentalism, and anti-war activism as part of being on the Left. But if there is one characteristic that is essential and overriding, it is the desire for greater economic equality. Where opinions differ is over how much equality and how it should be achieved. Opinions further divide over the relation between economic issues and others. Marx took all social problems to be at root economic. Many current Leftists think that, contrary to Marx, racism and sexism have a life of their own; they are independent of economic factors and must be separately combated.

The spectrum of opinion on the Right is perhaps even wider. Notions of equality for its own sake–economic or social–are typically shunned. Freedom is commonly stressed. Often tradition as a source of wisdom is upheld; this is especially true of social conservatives. On the other hand, champions of "unfettered free enterprise" often turn out to be social revolutionaries, which in established political parties can lead to serious conflicts with their social conservative allies.

Very often the local social situation can shape the form of Left or Right opinion. The Left in the United States and in most of the world, for instance, is anti-nationalist, taking the view that nationalism is an anti-progressive force. In Canada, on the other hand, the Left is highly nationalistic, adopting the view that it must protect Canada from American encroachment, which will lead to the undermining of its progressive institutions such as national health care. Multiculturalism, to cite another example, is a prominent cause for the Left in the United States. However, the Left in Canada is somewhat ambivalent, since it puts all cultures on a par, thereby undermining the special claim of Quebec (which the Canadian Left supports) in their struggle to maintain the French language and a separate culture in a sea of English-speakers.

This gives some idea of the spectrum of opinion and some idea of why we can't be precise in characterizing Left and Right. But if we say that for the Left, social and economic equality are paramount, we won't go too far wrong. And how social and economic equality relate to attitudes to science is the fundamental question.

Now to Snow's famous polemic.

Two Cultures and the Science Wars

The literati were much irritated by C. P. Snow's *Two Cultures and the Scientific Revolution* (1959/63). In this now classic work, Snow asserted that there are two cultures, in England and indeed throughout much of the world, that have little or nothing to say to one another, and that this is an unfortunate and even dangerous turn. One culture is scientific, the other literary. The former find Dickens obscure, the latter glaze over when asked about the second law of thermodynamics. In this little book, both sides seem at first to come in for equal scolding from Snow, the scientist turned novelist. But no, the literati are the worse offenders. And as we continue through his work, we find that "literary culture" (the culture of the literary intellectuals) gives way to "traditional culture" (the stuff of most ordinary folk), and this comes in for much general criticism. Snow ends with a call for major educational reform with much greater emphasis on the sciences in his favored scheme of things. But this is to put it in the mildest way.

When mentioned today, Snow's term, "two cultures," invokes the gulf of unintelligibility between the humanities and the sciences. Even Jeremy Bernstein gets it wrong and he's both a scientist and a writer, like Snow himself. "I am . . . unimpressed with C. P. Snow's discussion of the 'the two cultures.' I wonder if anyone still reads him." Evidently not Bernstein, since he adds: "what Snow is worried about most when it comes to the two cultures seems to be High Table conversation" (1996, 12). But there is much more in Snow's rambling and somewhat disjointed essay than fear of a chasm of incomprehension over dinner. Especially interesting and important are his sporadic comments on the politics of the two sides. Snow identifies the scientific culture with the Left (at least statistically) and the literati with the Right. He confidently asserts that scientific culture could deal much better with serious social problems, and he even makes the inflammatory claim that the major literary figures of the early twentieth century–Yeats, Pound, and Wyndham Lewis–brought us a little closer to Auschwitz.

What did Snow care about? He cared about *who does rule* and *who should rule.* The literati, in his view, were part of the existing ruling class, a pack of muddle-headed right-wingers who have not been able to deliver a decent living to the great mass of humanity. The scientific Left could do very much better and should be given the chance.

Scientists liked the piece, but the literati were incensed. Perhaps the most famous reply came from F. R. Leavis, the Cambridge literary critic, who mocked and denounced Snow on many scores in his *Two Cultures? The Significance of C. P. Snow* (1962). Leavis refrained from calling Snow a bad novelist only because he refused to call him a novelist at all.

Some defenders of literary culture, such as Michael Yudkin, were more temperate. But his remarks in defense of literary and traditional culture actually shed more light on the depth of the divide and even reinforce it. "To read Dickens, or to hear Mozart, or to see a Titian can be in itself a rewarding activity," says Yudkin, "but to find out what is meant by acceleration is to gain a piece of factual information which in itself has no value" (Leavis and Yudkin 1962, 35).

This is a stunning remark. The scientific concept of acceleration was hard won; to understand it is to make a crucial distinction between *velocity* and *change of velocity;* it is to understand *force,* and to grasp the connection between any kind of change and the mathematical notions of derivative and integral, so brilliantly put together by Newton and Leibniz. To put it on a par with, say, the fact that I keep the forks in the drawer on the left is to miss the point of scientific culture. It's a bit like saying: "Well, perhaps Mozart makes the cows give more milk; but besides that, so what?" It is little wonder that scientists (regardless of politics) are often contemptuous of nonscientist commentators on science.

The battle long ago over Snow's *Two Cultures* has much in common with the current science wars. The differences, however, are more important. What they seem to have in common is a dispute between scientific and the literary types. The warring parties are similar. But that's the end of it. In the current round of fighting, it's (1) the objectivity of science, rather than its cultural importance, that is principally at issue (though its cultural importance is certainly a factor). In Snow's day, this was not at issue. As Yudkin might have put it: science deals with mere facts. Now the focus is different; current science studies commentators are more likely to say that science is as nonfactual as any other part of general culture; its so-called facts are social constructions. (2) Snow was fighting for underdog science, whereas today science is anything but that. It is the literary side who feel under siege from a dominant techno-science culture. (3) Snow took his scientists to be predominantly on the Left and the literati on the Right. Yet, in much of the current science wars commentary, the very opposite is assumed. Gross and Levitt, for example, in *Higher Superstition,* take the opponents of science to be, in their words, the "academic left."

I could go on listing similarities and differences. And not just between Snow's "two cultures" and the rival sides in the current science wars; we could find lots of interesting similarities and contrasts with much earlier clashes such as the Romantic rejection of the Enlightenment. Instead, let's turn to Alan Sokal's now famous prank.

SOKAL'S HOAX

Is the "Sokal Affair" old news? Its persistence is remarkable. Books and articles continue to appear.[1] Ian Hacking confesses: "a few days after the story broke in May of 1996, I said that Sokal's hoax had now had its fifteen minutes of fame. How wrong I was" (1999, 3). The infamous hoax may be history, but how much has the culture absorbed? Let's start with a quick quiz. One of the following statements is generally accepted by physicists. The other is a deliberate piece of mushy nonsense. Which is which?

a. Henceforth space by itself, and time by itself, are doomed to fade away into mere shadows, and only a kind of union of the two will preserve an independent reality.
b. The first homology group of the sphere is trivial, while those of the other surfaces are profound; and this homology is linked with the connectedness or disconnectedness of the surface after one or more cuts. Furthermore, there is an intimate connection between the external structure of the physical world and its inner psychological representation qua knot theory.

Did you have trouble picking out the nonsense? So did the luckless editors of *Social Text,* who are still wiping egg off their faces. They were taken in by Alan Sokal, a working physicist with professed left-wing sympathies. He submitted–and they unwittingly accepted–his parody, "Transgressing the Boundaries: Toward a Transformative Hermeneutics of Quantum Gravity" (Sokal 1996a), a concoction of cleverly contrived gibberish written in the worst postmodern jargon. Within days Sokal revealed his hoax in *Lingua Franca* (Sokal 1996b), saying he was not so much trying to defend science from its empty-headed critics as he was hoping to rescue the political Left from a disastrous form of thinking.

Agitation and anger abound. Diverse issues have been lumped together and the active participants polarized into two camps to fight *the science wars,* as it has come to be called. War, we are told, is politics by other means. This is especially true of the science wars, where social goals loom large. Shifting aims and unnatural alliances make war-watching intriguing but difficult to follow. Some of the concerns, only an academic could fuss over; others are central to how we live our lives. But more of this in a moment. Now, back to Sokal's hoax.

Derrida, perhaps the most prominent postmodern of them all, offers us this piece of fluff:

> The Einstein constant is not a constant, is not a center. It is the very concept of variability–it is, finally, the concept of the game. In other words, it is not the concept of a some*thing*–of a center starting from which an observer could master the field–but the very concept of the game.

With tongue firmly in cheek, Sokal sympathetically elaborates:

> In mathematical terms, Derrida's observation relates to the invariance of the Einstein field equation $G_{\mu\nu} = 8\pi G T_{\mu\nu}$ under nonlinear space-time diffeomorphisms . . . In this way the infinite-dimensional invariance group erodes the distinction between observer and observed; the π of Euclid and the G of Newton, formerly thought to be constant and universal, are now perceived in their ineluctable historicity; and the putative observer becomes fatally de-centred, disconnected from any epistemic link to a space-time point that can no longer be defined by geometry alone. (1996a, 222)

And when Jacques Lacan incorporates the Möbius strip into psychoanalysis, Sokal amplifies approvingly with hokey remarks about homology groups. By the way, this particular tidbit of topological tomfoolery was the deliberate nonsense in the quiz above. It was the second of the two quotes; the first passage about space and time giving way to space-time is a remark (also quoted by Sokal) from Herman Minkowski's famous 1908 article that reconceptualized special relativity and set the framework for general relativity.

Of course, Sokal's musings are all bluff and bluster. But who, precisely, was the target of this parody? In his hoax, Sokal criticizes scientists who

> cling to the dogma imposed by the long post-Enlightenment hegemony over the Western intellectual outlook, which can be summarized briefly as follows: that there exists an external world, whose properties are independent of any human being and indeed of humanity as a whole; that these properties are encoded in "eternal" physical laws; and that human beings can obtain reliable, albeit imperfect and tentative, knowledge of these laws by

> hewing to the "objective" procedures and epistemological strictures prescribed by the (so-called) scientific method. (1996a, 217)

By pretending to attack, Sokal is actually defending such a view, an outlook more or less shared by working scientists and presumably by Sokal himself when he's not posing as a pomo (as postmoderns are often called). Crudely, we can put it like this: *There is a way things are, and scientists are trying to figure it out; they have a variety of (fallible) techniques for doing so and thus far have been quite successful.* Let us call such a view *scientific orthodoxy.* I'm happy to leave it to readers to fill in their favorite version of the details, but if pressed for my version of scientific orthodoxy I would include the following:

a. There is a world in which there are objects, processes, and properties that are independent of us and our beliefs about them. Any statement we make about them is false or is true (or at least approximately so). Of course, we may never know which.
b. The *aim* of science is to give true descriptions of reality. Science can have other aims as well (usually associated with technology), but truth is the chief aim of pure science.
c. We have a variety of tools and techniques (observation, logic, statistical inference) for learning how things are. These methods have developed from earlier methods and very likely will themselves be developed further.
d. Such methods are fallible; they may lead us astray. Nevertheless, science has made remarkable progress so far. It is reasonable to continue to use these methods (and to continue refining them) in the belief that they are the most reliable source of information about nature.

This cluster of views is just common-sense realism. (A more detailed discussion will be found in Chapter 4.) Common sense is sometimes misleading, however, and in this case, say the so-called social constructivists whom Sokal is parodying, it is profoundly wrong. But is it? One of the many aims of this book is to give a reasonable answer.

A word or two about "science" is also in order. I'd prefer to leave the term slightly vague, but not too ambiguous. The principal meaning of "science" is that body of current theories that purport to truly describe the world (or at least to systematize our experience). This should be contrasted with technology and applied science, the attempt to control and manipulate the world in the pursuit of our practical goals. "Science" also means the institution of science, the vast collection of universities, research establishments, and government agencies. It's important to separate these different meanings conceptually, even though they are all involved in many of the crucial policy questions that deeply affect our lives.

Let's return to Sokal again. Why did he perpetrate his hoax? In his exposé Sokal declares a political purpose. "I'm an unabashed Old Leftist who never quite understood how deconstruction was supposed to help the working class [or promote social justice in general]" (Sokal 1996c). Far from believing that science needs saving from the postmodern crowd, Sokal holds that left-wing politics needs rescuing from idiotic thinking. Like many before him, Sokal holds that relativism, irrationalism, and downright sloppy reasoning undermine progressive political aims. The Left doesn't have the money or the guns to get its way. Clear thinking is the Left's best weapon. And only clear idiocy would abandon it.

To a large extent, then, the real issue between Sokal and his postmodern target is not, "How should we think about quantum gravity?," but rather, "How can we best change society?" Of course, Sokal is not the first to raise such issues, nor will he be the last–but he was certainly the most dramatic.

So how did the science wars get started? Right from the beginning constructivist views have been criticized. A particularly bitter phase began with the publication of *The Higher Superstition* by Paul Gross and Norman Levitt (1994). The subtitle of their polemic is *The Academic Left and Its Quarrels with Science.* It suggests a clear target, but the term "academic Left" is hopelessly vague. Gross and Levitt use it "to designate those people whose doctrinal idiosyncrasies sustain the misreadings of science, its methods, and its conceptual foundations that have generated what nowadays passes for a politically progressive critique of it" (1994, 9).

To define the academic Left as those who "misread" is hardly a generous characterization. It is seldom a wise strategy to characterize one's foe as *wrong by definition*–victory is too easy and too short-lived. Instead of defining the academic Left as those who misread science because of their doctrinal idiosyncrasies, it would be much better to use the term "left-wing critics of science" (defined as those who criticize science from a left-wing point of view) and to save the term "academic Left" for what it really represents, namely that set of academics who have left-wing views, allowing the obvious fact that some are and some are not critics of science. Sokal, like many others, is a member of the academic Left, but is not a left-wing critic of science. Parallel terms will also be useful: "academic Right" is similarly neutral in attitudes toward science, whereas "right-wing critics of science" would include, for example, many religious detractors of Darwin. With this apparently petty–but actually important–terminological point out of the way, we can get back to the main story.

Higher Superstition caused quite a stir when it appeared in 1994. Its authors gave several popular and well-attended talks. Some critics, like Andrew Ross (an editor of *Social Text*), saw it as part of a general backlash, an attack on progressive movements in race relations, in gender relations, in the environment, and so on. Thus was born the special issue of *Social Text* with the title "Science Wars." To some extent the prominent use of the term stems from this, and all sides have found it convenient, since the term is fairly neutral (though there are grumblers–including Sokal himself). The relative neutrality of "science wars" is quite unlike "political correctness," which started out as playful, leftist, and feminist self-mockery, but in the hands of their opponents quickly became a term of sneering derision.

In his introduction to the "Science Wars" issue of *Social Text,* Ross uses such expressions as "undemocratic" in characterizing science. This is guaranteed to polarize. Scientists will be deeply offended at being so labeled. Or, perhaps, they will respond by saying that, of course, science isn't democratic–we discover the truth, not vote on it. But when Ross discusses a different point, the true character of his attack on "anti-democratic" science is clarified. He is worried, for example, about the relations between "remote" scientific experts and the local worker.

> The unjustified conferral of expertise on the scientist's knowledge of, say, chemical materials, and not the worker's or the

> farmer's experience with such materials, is an abuse of power that will not be opposed or altered simply by demonstrating the socially constructed nature of the scientist's knowledge. That may help to demystify, but it must be joined by insistence on methodological reform–to involve the local experience of users in the research process from the outset and to ensure that the process is shaped less by a manufacturer's interests than by the needs of communities affected by the product. This is the way that leads from cultural relativism to social rationality. (Ross 1996, 4)

There's much crammed into this passage, and with much of it any champion of scientific orthodoxy could agree. The remarks about social constructivism would be rejected, but no scientific rationalist thinks that a local worker's or a local farmer's experience is irrelevant. On the contrary, rational belief should be decided by evidence, and the local workers' experiences are part of the total evidence to be considered. Moreover, it comes as no shock to the most zealous advocate of orthodox science that a farmer might have a different interest than a chemical manufacturer.

Ross believes there is very much more going on. He sees the science wars as a new front in the "holy Culture Wars. Seeking explanations for their loss of standing in the public eye and the decline in funding from the public purse, conservatives in science have joined the backlash against the usual suspects–pinkos, feminists, and multiculturalists of all stripes" (Ross 1996, 6). To what extent, if any, is he correct?

The Ensuing Skirmishes

Sokal published his hoax article in the "Science Wars" issue of *Social Text* in the spring of 1996, and immediately followed with his exposé, "A Physicist Experiments with Cultural Studies," in the May/June issue of *Lingua Franca.* The subsequent exchanges in the popular media, the academic press, and especially on the Internet were fast, furious, and fascinating.

The *New York Times* gave it prominent coverage right from the start. In the May 18, 1996 issue, Janny Scott's front-page article described

Sokal's prank and declared it "one more skirmish in the culture wars, the battle over multiculturalism and college curriculums and whether there is a single objective truth or just many differing points of view" (Scott 1996).

"Is the jargon-infested writing in postmodern academic journals as stupid and empty as it appears? Can people teaching cultural studies get away with absolute nonsense? Sokal devised an experiment, and came up with an answer: Yes, in both cases." So wrote Robert Fulford in the *Globe & Mail* (June 5, 1996); he fears Derrida's influence is here to stay, but is at least grateful for the "glimmer of hope" provided by Sokal.

A "Pomolotov Cocktail" is what Katha Pollitt called it in the left-liberal magazine *The Nation* (June 10, 1996). "Sokal's demonstration of the high hot-air quotient in cultural studies–how it combines covert slavishness to authority with the most outlandish radical posturing–is, if anything," Pollitt writes, "long overdue" (1996, 9). She does note a downside, however. "Unfortunately, another effect of his prank will be to feed the anti-intellectualism of the media and the public." She deplores the fact that brilliant sociological work will be lumped in with the rubbish, and that respectable people "will have to suffer, for a while, the slings and arrows of journalists like the *Times*'s Janny Scott, who thinks 'epistemological' is a funny word" (ibid.). As for the work of the postmodern crowd, "What results," says Pollitt, "is a pseudo-politics, in which everything is claimed in the name of revolution and democracy and equality and anti-authoritarianism, and nothing is risked, nothing, except maybe a bit of harmless cross-dressing, is even expected to happen outside the classroom" (ibid.). In similar spirit Gary Kamiya (1996) referred to the "Transformative Hermeneutics of Total Bullshit"–not wanting to put too fine a point on it.

Even the *Village Voice* got into the act with a piece by Ellen Willis called "My Sokaled Life," which correctly predicted that "We can now look forward to months of having the Sokal affair trotted out as definite proof that radical criticism of science, and indeed the entire enterprise of cultural studies, boils down to mindless, fraudulent gibberish" (Willis 1996).

Martin Gardner–everyone's favorite polymath–was delighted with the hoax in his "Notes of a Fringe-Watcher" column in the *Skeptical In-*

quirer. As well as making the anti-constructivist points that are by now familiar, Gardner takes a swipe at Thomas Kuhn's *Structure of Scientific Revolutions,* which "has been responsible for much postmodern mischief" (1996, 16). We'll be looking at Kuhn's views below.

There were interesting reactions to the Sokal affair throughout much of the world. After the United States, reactions may have been strongest in France. Most of Sokal's prominent targets were French. With a Belgian colleague, physicist Jean Bricmont, Sokal wrote *Impostures Intellectuelles* (1997) (in the United Kingdom: *Intellectual Impostures;* in the United States and Canada: *Fashionable Nonsense* (1998)), which is a detailed critique of the misuse and abuse of science in the work of several leading French intellectuals, including Jacques Lacan, Julia Kristeva, Luce Irigaray, Jean Baudrillard, Bruno Latour, and Gilles Deleuze. When word of its impending publication got out, the pages of *Le Monde* and other French dailies were filled with angry exchanges. Some, including Latour and Kristeva, called Sokal anti-French and even anti-European; others, including many French, defended Sokal and heaped additional scorn on the postmoderns. There is no doubt that current French intellectuals must be feeling particularly put upon, especially when they see major foreign newspapers such as the *Guardian Weekly* asking in an eye-popping headline: "Is Modern French Philosophy Just a Load of Pseudo-Scientific Claptrap?" (Henley 1997).

Not all interventions have been quite so partisan. The very prestigious British journal, *Nature,* called for "respect and rigour" (in the Jan. 30, 1997 issue). "The stakes on both sides are high. On the one hand, some scientists believe that they are fighting for the intellectual and social credibility of an enterprise that remains essential for human wellbeing. On the other, many social scientists argue equally convincingly that only a deep understanding of science as a social (as well as intellectual) process will enable us to strengthen the bridge between the worlds of science and politics that is essential if this well-being is to be achieved" (ibid., 373). This sounds the sweet voice of reason and we can certainly applaud *Nature*'s good will, but without some guide to which social constructivist claims are, and which are not, wrong-headed, the call for "respect and rigour" is rather empty.

Speaking of conciliatory, that seems to have been the prevailing—though I must say, unexpected—attitude at the PSA (the biannual meeting of the Philosophy of Science Association held in November 1996 in Cleveland). One need only peruse the pages of the journal, *Philosophy of Science,* to realize that the Philosophy of Science Association is no happy home for social constructivists. Sokal was an invited speaker; he recounted his hoax and drew the morals he has often drawn before and since. A number of very prominent philosophers of science (Arthur Fine, Philip Kitcher, and Ron Giere) spoke from the floor during the question period and expressed an ambivalence about Sokal's actions,[2] which I might summarize as "Yes, the hoax was funny, and yes, you (Sokal) were correct about the silliness of those you lampooned–but we really wish you hadn't done it. There was a debate going on, and now, thanks to your hoax, things have become so polarized that further discussion is greatly hampered."

If anything, the opposite has occurred. The Sokal affair has brought the science wars to a head. I find that an increasing number of people who would cheerfully characterize themselves as holding anti-postmodern, pro-orthodox views on science are starting to ask serious questions about specific social constructivist doctrines. Some social constructivists are at pains to distance themselves from postmodern critiques. And Andrew Ross, in a recent opinion piece, admits that the oft-repeated charge that postmodern language is hopelessly obscure is a fair complaint.

There are signs of conciliation all about. Conciliation is not always a real solution, however, and shouldn't be turned into a fetish. Some differences of opinion are settled by coming to a consensus after compromise. Others are settled by having clear winners and clear losers, the latter withdrawing into the woodwork to labor in wretched isolation. Since much of the science wars is about how we live, the outcome will affect huge numbers of people who are not intellectually directly involved in the debate itself. Whether a handful of participating academics feel victorious, humiliated, or just manage to save face is of no real importance. This is not like working out a compromise on parking privileges. Getting things right is of paramount importance for society in general.

The Political Side

Were those who worried about the political Right being cheered by Sokal's antics merely paranoid? Not at all. Those who saw Sokal's hoax as a right-wing backlash must have felt vindicated when prominent arch-conservatives quickly got into the act. Rush Limbaugh, on his radio program (May 22, 1996), declared that the gibberish of Stanley Fish (a leading postmodern literary critic and social commentator) is indistinguishable from Sokal's deliberate gobbledygook, and that academics like Fish are effete elitists, living in a different world. In the same broadcast he went on to attack the "pseudo-history" produced in African studies programs. George Will (in his May 30, 1996 column) remarked that Sokal's hoax reveals the "gaudy silliness of some academics." He was quick to denounce "using higher education's curricula to dole out reparations to 'underrepresented cultures'" (Will 1996).

"Cultural studies" was again the focus of animus in Roger Kimball's "'Diversity', 'Cultural Studies' & Other Mistakes" (in *The New Criterion,* May 1996). "The irony, of course, is that many of the students who agitate against the 'Eurocentric' curriculum at Columbia and elsewhere are only present at the university in the first place," says Kimball, "because of the discriminatory practice of what is euphemistically called 'affirmative action,' a.k.a. preferential treatment. But once admitted, it turns out that what many such students want is not an education but ideological training designed to confirm their coveted status as 'victims'" (1996, 4). *Social Text*'s editor, Andrew Ross, is singled out for special mention by Kimball, who scruples not at the *ad hominem:* "A Scot by birth, Professor Ross has parlayed an accent (proletarian chic), adolescent attire and intellectual interests, and large dollops of Marxist rhetoric into an amazingly successful academic career" (ibid., 5). Kimball's delight with the Sokal hoax was expressed later in *The Wall Street Journal* (May 29, 1996), where he urged "deans and presidents, parents and alumni, legislators and trustees, to take a hard look at the politicized nonsense they have been conned into subsidizing."

The political Right was gleeful, but many on the Left greeted Sokal with equal enthusiasm. Writing in the *Los Angeles Times,* historian Ruth Rosen wholly endorsed the prank. "Sokal's spoof exposed the hypocrisy practiced by these so-called cultural revolutionaries. They claim to be democratizing thought, but they purposely write in tongues for an initiated elite. They claim that their work is transformative and subversive, but they focus obsessively on the linguistic and social construction of human consciousness, not on the hard reality of people's lives" (Rosen 1996). She knows there is a price to pay for this sort of thing, but it is worth it: "Yes, I know that the conservative right may use Sokal's parody to further attack 'tenured radicals.' But if the progressive left is to survive and be credible, it must withstand the glare of public scrutiny and be worthy of people's respect" (ibid.).

Michael Albert of *Z Magazine*–a leading left-wing periodical that publishes many of Noam Chomsky's political pieces–applauded Sokal and his motives. The Left has no weapons of its own but logic and rationality, he notes. Why abandon them? "Elites not only have the guns and money," says Albert, but if the postmoderns have their way, then "we let them have 'truth' too. It would be hard to conceive a more self-defeating stance" (Aug. 1996, *Z-Net*).

> To critique bad science and scientists is part of understanding the world to make it better. To suggest methods people could use to avoid excessive reductionism, to guard against exaggerating the scope of scientific insights into domains where it is inapplicable (and there are many), or to ward off sexist, racist, and classist biases is a useful way to aid sincere scientists (and left political activists). But to critique reason and logic as being at the root of science's many evils is wrong and has no role in making the world better. And that is what rejecting the scientific method amounts to. (Albert, Aug. 1996, Z-Net)

The Counter-Attack

A few days after the *New York Times* reported the hoax, Stanley Fish launched a counter-attack in the paper's editorial pages. Fish is a leading figure in "cultural studies." His "Professor Sokal's Bad Joke" quickly became a focal piece. In it he decried the "corrosive effects of an academic hoax." More bizarre–if not disingenuous–was Fish's claim that social constructivists have actually honored scientists and not denied the reality of the world or its properties. "They [social constructivists] just maintain and demonstrate that the nature of scientific procedure is a question continually debated in its own precincts. What results is an incredibly rich story, full of honor for scientists." He then drew a curious analogy designed to show that "real" and "socially constructed" are perfectly compatible. In baseball, "strikes" and "balls" are clearly social constructions; that is, they are human creations, not found in nature. However, once the constructions are made, a particular pitch really is a strike or a ball. "The facts yielded by both [baseball and science] will be social constructions and be real."

One wonders how good an analogy this is. The rules of a game are rather obvious social constructions, but are molecules and the moon made the same way? Silver dollars are social constructions, but is silver? Fish wants the definition of a strike to be a social construction but allows it to be an objective fact that the trajectory of a ball did or did not pass through a particular region of spacetime. Has any die-hard scientific realist ever claimed otherwise? In many obvious senses silver dollars are similarly social constructions; but if silver itself isn't, then chemists have nothing to fear from Fish. His analogy is deceptive. It seems to work because he claims (1) strikes are invented, not discovered, and (2) trajectories are discovered, not invented. Most people would agree. The trouble is that no self-respecting social constructivist would concur with (2). Fish doesn't comprehend (or deliberately obscures) how far-reaching are the views of those he defends.

Scientists Have Their Say

Steven Weinberg is one of the most public of contemporary scientists. As well as being a Nobel Prize–winner for his work in high-energy physics, he writes some of the best popular science available (e.g., *The First Three Minutes* and *Dreams of a Final Theory*). He'll have no truck nor trade with constructivists, as he made perfectly clear in his *New York Review of Books* piece, "Sokal's Hoax" (August 8, 1996). Weinberg speaks for many in the scientific community when he allows that the *language* of science is a social construction, but nevertheless insists on the objectivity of physical laws. "To put it another way, if we ever discovered intelligent creatures on some distant planet and translate their scientific works, we will find that we and they have discovered the same laws" (1996, 14).

It's hard to imagine anything less likely to be true. For one thing, we ourselves have not always believed the same laws. The only way Martians, say, and ourselves would have the same beliefs is by having the same *history* of beliefs–a long Ptolemaic stage of astronomy, followed by a briefer Copernican/Keplerian/Newtonian interval, then on to general relativity, and so forth. Not only would our histories have to match, but the Martians had better be at the same stage as we are when we meet, otherwise it could be their Priestley counterpart meeting our Planck. Perhaps Weinberg would go to the wall for phlogiston, but not the rest of us. Intellectual determinism of this order is preposterous.

If faced with his choice: believe all intelligent creatures have discovered the same laws *or* believe social constructivism, it's easy to see why many thoughtful people opt for the latter. There are so many accidents, so many little things that affect the course of intellectual life, that it is extremely unlikely that two unconnected intellectual communities would have identical histories. Yet this is what Weinberg's view implies.

Social constructivists have often suggested exactly the same stark and implausible alternatives as the ones Weinberg offers. Unlike him, they choose to reject the objectivity of laws and to accept some form of constructivism as being the far more sensible view. But the dichotomy–certainty vs. construction–is a false one. Scientific objectivity is compatible with a high degree of fallibility. In their day phlogiston, caloric, and the aether were the stuff of excellent theories. Sensible people believed them on the best of grounds. But there's no reason to think that Martian physicists would ever pass through similar stages in their intellectual history.

Weinberg is well-aware of the "whiggism"[3] involved in his view, but he boldly clings to the present. "A historian of science who ignores our present scientific knowledge seems to me like a historian of U.S. military intelligence in the Civil War who tells the story of McClellan's hesitations in the face of what McClellan thought were overwhelming Confederate forces without taking into account our present knowledge that McClellan was wrong" (ibid., 15).

How does this help? To be "whiggish" is to be overly present-minded–a sin, but a subtle one. Rather than try to understand the past in its own terms, whiggish history imposes current concerns and concepts on the past; it results in a false account of what actually happened. A full historical account will, of course, include the fact that McClellan was wrong. If the aim is to explain why McClellan hesitated, then the truth about the relative strengths of opposing forces–which was only discovered later–has nothing to do with it. The thing to be explained is McClellan's hesitation. His *belief* that opposing forces were overwhelming explains that hesitation. This belief in turn is explained on the basis of the *evidence available to him at the time.* The truth itself was not directly available, so it played no role in his beliefs. Of course, evidence is intimately related to the way things are–that is, to the truth–but the relation is not transparent. If it were, science wouldn't be so difficult. Weinberg himself didn't win a Nobel Prize for spotting the obvious.

In his earlier book, *Dreams of a Final Theory,* Weinberg attacks all philosophers of science, not just social constructivists. "The insights of philosophers have occasionally benefited physicists, but generally in a negative fashion–by protecting them from the preconceptions of other philosophers" (1992, 166). Saving us from bad philosophy of science is the one area in which Weinberg allows philosophers to excel. When they read Weinberg's philosophical remarks on truth, most philosophers of science, I dare say, would be delighted to exercise this skill for his benefit.

The Wider Issues

The science wars are very much wider than Sokal and his specific targets. The social constructivist branch of so-called science studies is huge and diverse. Some of the people lampooned by Sokal (e.g., Derrida, Lacan) are not in science studies proper, but rather concern themselves with more general cultural matters that are rather unrelated to science. Though right to lampoon them, Sokal did pick easy targets. Let's cast our net a little wider. We can start by asking: Who's involved? Why should we care? What are the main battle lines?

In some ways the fight is quite old. Protagoras championed a kind of relativism 2,500 years ago when he said, "Man is the measure of all things." Plato took up the challenge and fought for objective knowledge. The Enlightenment with its emphasis on progress through rationality was no sooner established in the eighteenth century than early in the nineteenth it faced the Romantic rebellion, which stressed feeling over intellect and emotion over rational inference. Much debate in more recent times has been stimulated by Karl Marx, though sometimes his writings pull in opposite directions. Marx sounds distinctly like a social constructivist when he famously declared: "The mode of production of material life conditions the general process of social political and intellectual life. It is not the consciousness of men that determines their existence, but their social existence that determines their consciousness" (1859, 20f). Yet Marx also thought that objective knowledge is possible; the constructive sentiment gives way to a sensible though subtle form of realism:

> With the change of economic foundation the entire immense superstructure is more or less rapidly transformed. In considering such transformations the distinction should always be made between the material transformations of the economic conditions or production which can be determined with the precision of natural science, and the legal, political, religious, aesthetic or philosophic–in short, ideological–forms in which men become conscious of this conflict and fight it out. (ibid., 21)

Current social constructivism has plenty of antecedents, but it is also reasonable to think of it as mainly a product of the past three decades. In the mid-1970s David Bloor (in Edinburgh) announced the *strong programme* in the sociology of knowledge. Why *strong?* It's in opposition to *weak* sociology of science; any account that focuses on institutions and various other social features of science but takes for granted that the *content* of science has nothing to do with sociology. By contrast, Bloor asserts that the very content of scientific theories is also to be understood in terms of social factors.

The importance of Bloor's point must be stressed, since a great deal of sociology of science is quite compatible with the epistemology of scientific orthodoxy but at the same time is potentially embarrassing to the orthodox. So-called weak sociology, for example, can ask: Why are there so few women physicists? Why do they feel they must sacrifice career or children, and can't (unlike their male colleagues) have both? However, weak sociology of science does not ask questions such as: Why do female physicists accept the standard model of elementary particles? Why do women believe that the trajectory of a cannonball is a parabola? The answer to these questions is "the available evidence" and it has nothing to do with their sex, nor with any other sociological factor. Bloor's strong programme will have none of this hands-off attitude.

He, too, will ask the background questions. But as likely as not, he will relate them to the content of the theory at hand.

Shortly after Bloor started to make his mark in science studies, Bruno Latour (a French philosopher and anthropologist) adopted the role of an "anthropologist in the lab." With Steve Woolgar he wrote up his experiences of an exotic tribe–a team of California biochemists–explaining their behavior in social, political, and economic terms. Meanwhile in France, Michel Foucault was claiming that *knowledge = power,* not in the sense that by having knowledge one has power (a sense made famous by Francis Bacon), but in the very different sense that having political power allows one to say what knowledge is and is not. Elsewhere, feminist critics of science such as Evelyn Fox Keller and Sandra Harding were claiming that science reflects the scientists who made it–males–and that women could and would make better science. Even this brief sketch indicates why their views might be thought controversial.

The Hoax Itself

Far too much has been made of the details of Sokal's hoax. The important–and very useful–thing about the hoax is that it brought matters to a head. Those who were stung by it complained that it was deeply dishonest and destroyed the trust that exists in academic life. Piffle. Defenders of the hoax say it clearly revealed social constructivism for the intellectual rubbish it is. Not so.

Though amused at the hoax when I first heard about it, I must confess to feeling a twinge. As a sometime editor,[4] I often commissioned book reviews and survey articles that I accepted on faith when they came in. I read them over, but (unlike research articles) they did not go out for so-called peer review. It would have been very easy for any of these authors to slip utter nonsense by me, at least if it were in an area of philosophy of science with which I am not familiar. I could have had these papers checked by experts, but it's so hard to find good, willing referees that I wouldn't dream of wasting them on book reviews or commission pieces. After all, I invited these people because they are known to be experts in the field. It was perfectly reasonable for the editors of *Social Text* to assume that Sokal knew his physics.

What about the fact that *Social Text* doesn't send out *any* papers for peer review? Perhaps book reviews and survey pieces are one thing, but research articles (which Sokal's hoax paper purported to be) are quite another. Here we need to understand–and defend, if necessary–the fact that some journals have a mission. Physics is a mature and well-established science. That's why *Physical Review* can referee all submissions in a straightforward way. But the "editorial collective" of *Social Text* have a particular social outlook that they want to explore and promote. In this they are not very different from a great many other academic journals. There are academic periodicals of sociobiology, for example, that will have rather limited peer review. They would not send out a submission which argued that racist behavior is genetically determined to Stephen J. Gould for review, though he's as knowledgeable as anyone on the topic. The only criticism considered is from those working within the paradigm (to use the well-worn Kuhnianism). And for that matter, what physics journal is likely to send something on the theory of relativity to Derrida to referee, even though (as we saw above), he's published on the topic?

Sokal's hoax proved little in itself. But it did do something of great (and I'll bet of enduring historical) importance. It raised a flag to rally around. Those who felt sympathetic both to the Left and to science suddenly felt like they had a bit of elbow room. Feeling cramped and uncomfortable with the pro-science Right or anti-science Left option, now they can march into political battle confidently armed with the most powerful weapon imaginable–rational thought.

Left vs. Left

Even supposing Sokal and other critics of social constructivism are correct, is their attack wise? Andrew Ross, the editor of *Social Text,* says that it is the great tragedy of the twentieth century that the "left eats the left." This has indeed been a great problem; all progressive movements are beset by doctrinal disputes.

The reaction of most people, including most scientists, is to say this is irrelevant: *The truth is the truth and that's all there is to it; political opinions have nothing to do with figuring out how the world works.* Social constructivists–perhaps with justice–find this view completely incredible. Social factors have been all too obvious in many of the results of so-called objective science; they have been uncovered as subtly operating in others; and they are suspected of working somehow or other in all the rest. But there's also a moral side to science studies—at least in some cases. There is a justifiable outrage at the current social situation. Rather than delve into details, let me just point out some simple, but stunning facts. In 1980 the richest 1 percent of Americans owned 22 percent of the wealth of that nation. That was at the start of the Thatcher-Reagan era. At the end of the Reagan and Bush administrations a few years later, the richest 1 percent owned 42 percent of the wealth. In Canada during the last decade the number of children living in poverty tripled to 1.5 million. Throughout much of the world, the trend has been similar—a massive shift of wealth from poor to rich.

Science, no doubt, has done much to create this wealth. What has science to do with its *distribution?* Have race and IQ studies played no role in this? Has sociobiology been neutral? Have the sciences of economics and sociology been uninvolved? Only the foolish could find this credible and only the wicked could assert it with apparent conviction.

On the other hand, what about high-energy physics, or the topology of differentiable manifolds? They would seem to be as unrelated as anything could be to the plight of the poor, neither helping nor harming. We need to sort what is politically related science from what is not.

At the beginning of the science wars, the common contrast was between the left-wing critics of science and conservative defenders. This is the way Gross and Levitt drew the lines of battle in their attack on the academic Left. They saw the battle as between the mush-minded Left on the one hand and people of good sense on the other.[5] And many of those they attacked implicitly agreed with this dichotomy. Numerous postmodern authors tend to view things the same way—a Left-Right battle—though they would characterize it as a fight between social progressives (i.e., themselves) and reactionaries who defend science as part of the general defense of the political status quo. In short, Gross and Levitt and many of their postmodern targets agree on this simple-minded but false dichotomy: *anti-science Left vs. pro-science Right.* The Sokal affair has made it perfectly plain that this is not the proper way to draw the lines of battle. This is why Sokal is so very important.

Politically conservative attitudes toward science are similarly divided, too. On the one hand, religious fundamentalists are at war with evolutionary biology. They correctly see that science is occasionally at odds with some of their deepest religious convictions, but they are in no mood to give up those convictions in the face of scientific evidence. On the other hand, some IQ researchers tell us that there is indeed a link between race and intelligence and that we'll just have to accept these facts, however unfortunate, since they were arrived at by a rigorous application of scientific method. And the consequence, in their view? The existing social structure is "natural," so don't try to change it. Being on the Right doesn't determine attitudes to science any more than being on the Left does.

The false dichotomy of anti-science Left vs. pro-science Right must be abandoned. Though still very simplistic, it would be much better to stake out four distinct positions–a pair of pro-science views and a pair of anti-science views. This is only tentative. Later we will have to make serious revisions to the picture, but it will serve us well for the moment. Recall that "scientific orthodoxy" was discussed above. I sketched my version; others might prefer something slightly different. A precise definition isn't necessary or even useful. Champions of "socialism" and "free enterprise" often disagree on the precise meaning of these terms, but in almost every context they have no trouble understanding what is meant and deciding which side they're on. The same can be said here about the "orthodox view of science."

	Political Left	*Political Right*
Hostile to orthodox views of science	Some social constructivists, postmoderns	Religious conservatives, anti-Darwinians
Friendly to orthodox views of science	Sokal, Chomsky, Gould, Lewontin, the Vienna Circle	Some sociobiologists, race and IQ theorists

Let me elaborate a bit on each of these positions. The top right view plays only a small role in the so-called science wars, but it is certainly an important view in the larger setting. It's worth mentioning here mainly for the sake of completeness and for a contrast with other views. Such an outlook is typified by the religious Right. The rejection of Darwin's theory of evolution and the upholding of the account of creation in the book of Genesis is perhaps most characteristic. More generally, there is strong opposition from this quarter to what is seen as the intrusion of science into aspects of life that should be governed by moral and religious concerns. The attitude says, for example: don't try to give scientific accounts of homosexual activity, or of juvenile delinquency, or of drug and alcohol dependency, or of teenage pregnancy, since doing so really amounts to making improper excuses for immoral behavior. We're in the domain of free choice, says the religious Right, where God's judgment, not science, belongs.

The top left view might go something like this: There is no such thing as the way the world is, a way that we could discover; rather, we impose frameworks on the world, and those frameworks serve various social and political interests. Current science is a framework that serves the rich at the expense of the working class, it serves men at the expense of women, the West at the expense of the third world, and it serves whites at the expense of people of color. Since there is nothing objective about science, we can adopt different theories that might better serve the oppressed. Existing science should be criticized for the ideology that it is and should be replaced by a science more friendly to the poor, more friendly to women, more friendly to the environment, and so on.

The bottom right is the view held by typical science conservatives. Its self-image is simply this: We're just discovering the facts. Some of the facts we've uncovered, say about race and IQ, might not be very pleasant, but that's not our fault. Sometimes the world just isn't the way we'd like it to be. Don't be angry with us, if you don't like what we have found–that's just shooting the messenger when you don't like the message.

In Gross and Levitt's terms, the top left position (which they call the academic Left) is battling it out with the bottom right (portrayed as ordinary scientific good sense). The Sokal affair illuminates one more position. The bottom left includes a number of pro-science people who are on the Left. The general idea is that objective science can be a powerful tool in progressive causes. It rejects any sort of all-encompassing social constructivism and instead maintains there is a world that exists independently of us, a world that science can describe and explain. When we understand how that world works, we will be in a better position to improve the lot of all people.

I said this fourfold categorization is still rather simplistic. How so? For one thing–to use a now contaminated expression–these are boundaries we often transgress. Many progressive champions of the orthodox view of science are highly critical of various parts of science. Thus, a completely orthodox but very careful statistical analysis might flush out fallacious reasoning about race and intelligence. Indeed, it is the ability to do this that makes so many on the Left critical of social constructivists–good science is the best weapon in the war on socially pernicious pseudo-science.

On the other hand, social constructivists are a very diverse group. It is a terrible misrepresentation to lump them together as political progressives and as hostile to the orthodox view of science. In North America social constructivists are perhaps more often on the Left than elsewhere, but it is certainly not essential to be so. Indeed, social constructivist views in the past have, more often than not, been associated with right-wing, even fascist views. Postmodern heroes include Nietzsche, who certainly couldn't be called "politically correct," and Heidegger, who joined the Nazi Party.

Moreover, many who follow Bloor's "science of science" think that their social constructivism is a mere consequence of their scientific approach to the study of science, with nothing particularly political about the approach itself. The spectrum of social constructivist opinion is enormous. Again, these are more things to be sorted out in the course of this book.

For my money the most interesting thing about the science wars is the fight for the Left. There is a wide cluster of social goals shared by those on the Left, but there is serious disagreement over the best way to attain them. This leads us to some crucial questions: What role, if any, does science play in creating, sustaining, or changing the social order? What role, if any, do social factors play in the production and maintenance of scientific theories? What view of science should the Left adopt to best promote its social goals? The balance of this book will be devoted to finding out.

2

The Scientific Experience

When scientists hear social constructivist remarks they reel. They stagger, they protest, they bluster, they even prepare hoaxes. Constructivist claims are utterly alien, they have no connection with how scientists see themselves. What sort of constructivist claims? Here's a sample. The first is from *Leviathan and the Air Pump,* a very influential and controversial work in the history of science by two of the most prominent practitioners of the sociology of knowledge, Steven Shapin and Simon Schaffer.

> As we come to recognize the conventional and artifactual status of our forms of knowing, we put ourselves in a position to realize that it is ourselves and not reality that is responsible for what we know. Knowledge as much as the [political] state, is the product of human actions. (Shapin and Schaffer 1985, 344)

A second example is from another famous and controversial work, *The Pasteurization of France,* by another leading constructivist, Bruno Latour.

> 'Reason' is applied to the work of allocating agreement and disagreement between words. It is a matter of taste and feeling, know-how and connoisseurship, class and status. We insult, pout, clench our fists, enthuse, spit, sigh, and dream. Who reasons? (Latour 1988, 179f)

Our third example is from another equally famous, influential, and controversial book, *The Science Question in Feminism,* by Sandra Harding, a leading feminist critic.

> [D]espite the deeply ingrained Western cultural belief in science's intrinsic progressiveness, science today serves primarily re-

> gressive social tendencies . . . its modes of defining research problems and designing experiments, its way of constructing and conferring meanings are not only sexist but also racist, classist, and culturally coercive. (Harding 1986, 9)

It's easy to imagine the visceral reaction most working scientists have to these sorts of comments, remarks that are at once alien and insulting and that, in their eyes, seem to stem as much from ignorance as from malice. If they only knew more science, the common reaction goes, these critics wouldn't utter such nonsense.

There is an interesting parallel with a feminist reaction to some forms of theorizing; it's only a rough one, but worth mentioning. Often feminists complain that various theories, especially in philosophy or the social sciences, fail to do justice to "women's experience." A particular ethical theory, for example, may make proposals about the nature of right and wrong and offer principles to guide our actions. Such a theory might stem, for example, from the problems two businessmen (both autonomous, rational agents) have negotiating a contract, and it might even provide brilliant methods of dealing fairly with those problems. The feminist objection is that theories such as this (taken to be general moral theories) do not take into account the fact that women spend much of their time raising children or looking after the old and infirm, and that consequently, much of women's moral decision-making involves dealing with those who are not autonomous, rational agents.[1]

This sort of objection to a moral theory is quite different from the direct attack, which attempts to locate fault with the theory in its own terms. The feminist objection is typically made not by pointing out this or that logical problem internal to the theory itself (though this can happen), but rather by relating the kinds of experiences that women face in day-to-day life and noting how the theory is largely irrelevant to much of this experience. Often this will be anecdotal, but with a few significant examples described in sufficient detail, even those of us who do not share the daily experiences of women can come to appreciate that such factors must be taken into account in a complete ethical theory.

Women are not alone in this regard. People of color often speak of the "black experience," immigrants of the "immigrant experience," and so on. As an argumentative technique it requires a measure of good will on the part of others–something not always forthcoming. It also suffers from a degree of vagueness and reliance on anecdotes–making the whole idea behind this approach easy to dismiss. Nevertheless, it is still powerful and useful, and in some fields such as ethics, its influence for the better has been keenly felt.

Scientists, I suspect, feel somewhat similar. When faced with social constructivist claims, they may try to meet them head on and–with only mixed results–to point out this or that specific fallacy. But for the most part, working scientists find sociological analyses of scientific knowledge far removed from anything like their normal experience. In the face of sociological analyses, working scientists are simply incredulous. So even if they don't know where to begin in offering effective criticisms, they feel so alienated from typical social constructivist work that they are likely to dismiss it completely. Of course, scientists are hardly social underdogs, which makes them quite unlike women and minorities. But when they offer their "experience" in the face of constructivist accounts of what they do, they are launching the same walk-a-mile-in-my-shoes rebuttal–*you wouldn't say that if you knew what I know.*

The only way to get a feel for the scientists' experience is to describe a number of examples. (I'll do this with a minimum of philosophical embellishment, but I should add that these examples are favorites with philosophers.) We might try to get at these by asking, What's so great about science? If we ask ordinary people why they find science so impressive, we usually get answers citing technological achievements: vaccinations are impressive, so are computers, so was the moon landing. If asked about the less inspiring side of science, we get answers concerning pollution or the lack of progress in fighting AIDS. Whether pro or con, the focus is on the technological. Working scientists themselves cast their votes elsewhere. The thing that inspires them about science is the brilliance of some theories in explaining the world. Science is great at telling us how things really are.

We are convinced that science can actually do this–but why? There are a variety of techniques–some quite sophisticated, others no more than common rules of thumb–that scientists employ in creating, modifying, and testing scientific theories. They go loosely by the name "scientific method." Science is great because it uses these established methods to produce its best results. Spelling out these methods in detail often leads to hot debates; nevertheless, something like the following are included in what I'll call "the scientific experience."

Examples, Examples, Examples

Novel Predictions

We're not much impressed by a new theory that predicts the sun will rise tomorrow, but the successful prediction of something quite unexpected is a triumph. What's the difference? With enough clever manipulation a theory can be made to imply or predict almost anything we want, so implying that the sun will rise may be merely the result of *ad hoc* fiddling. When it correctly implies something not already known to be true, however, then the accusation that the fix is in cannot arise; it seems as if the theory is genuinely on the right track.

One nice example of a prediction of the unexpected comes from plate tectonics.[2] Anyone who has looked at a map is struck by the way South America "fits" Africa. Were they once united but moved apart? Early in this century the prevailing view said no. The earth was molten and then cooled, according to received opinion. A crust formed, but as it cooled further the crust contracted and cracked. The continents did not move apart; they shrunk after cracking, but otherwise have not changed position. After all, how could they? There seems to be no force, no mechanism, that could push them about.

In the early decades of the twentieth century, Alfred Wegener held that early cooling resulted in a single, huge land mass (Pangea, as he called it), which subsequently broke up into continents. These continents, he maintained, have been slowly drifting around. His theory explains several things besides the apparent fit of various coastlines. He could also account, for example, for the similarities of plant and animal species in South America and Africa. In addition, the westward drift of North and South America accounted for the presence of mountain ranges running along the West Coast.

In spite of this, there was little or no sympathy for the drift hypothesis until the 1960s. After all, without a mechanism for moving the continents, drift seemed out of the question. Oceanographers in the 1950s changed all that: they discovered ridges running along the ocean floor, including one that runs up and down the mid-Atlantic. Notable features of a ridge include warmer water near the ridge and alternating directions in the strips of magnetic materials that lie parallel to the ridge. In 1960 Harry Hess speculated that convection currents of molten material under the ocean floor cause the ridges by rising and then spreading out. (The floor would reenter the earth's core in a trench that might be thousands of miles away.) This provided the mechanism of drift, since the continents simply ride on the expanding ocean floor. An interesting idea, no doubt, but the Hess view was still just speculation. It was the stunning confirmation of a surprising prediction that won the geological community over to the drift theory.

In 1963 Lawrence Morley and Fred Vine independently pointed out that the drift hypothesis had consequences involving the strips of magnetized material. When a magnetizable material is in a molten state, the molecules line up with the Earth's magnetic field. When the substance solidifies, the molecules maintain their orientation, providing us with a record of the past. It is known that periodically Earth's magnetic field has changed directions, suddenly flipping north and south poles. So, if the Hess conjecture is correct, then the alternating strips are alternating in their direction of magnetization because they solidified at different times (see Figure 1).

Similar alternations can be found in vertical strata. Here different layers are laid down at different times, later on top of earlier. The striking prediction made by Morley and Vine was that the strata on land and the strips parallel to the ridge should be perfectly matched. Oceanographic work done in 1966 established that the ocean floor patterns of magnetic materials did indeed match the pattern of magnetic reversals of land-based materials.

Quite aside from the theory that predicted it, no one would ever expect these patterns to match–it would be a fantastic coincidence. But, of course, no one thought this a mere bit of remarkable luck. The novel prediction made by the drift theory strongly suggests that the theory must be correct, or at least on the right track.

We can extract a principle of sorts from this example and others like it.

> *Novel predictions.* Suppose a theory T has a testable consequence C that, on independent checking, turns out to be correct. And

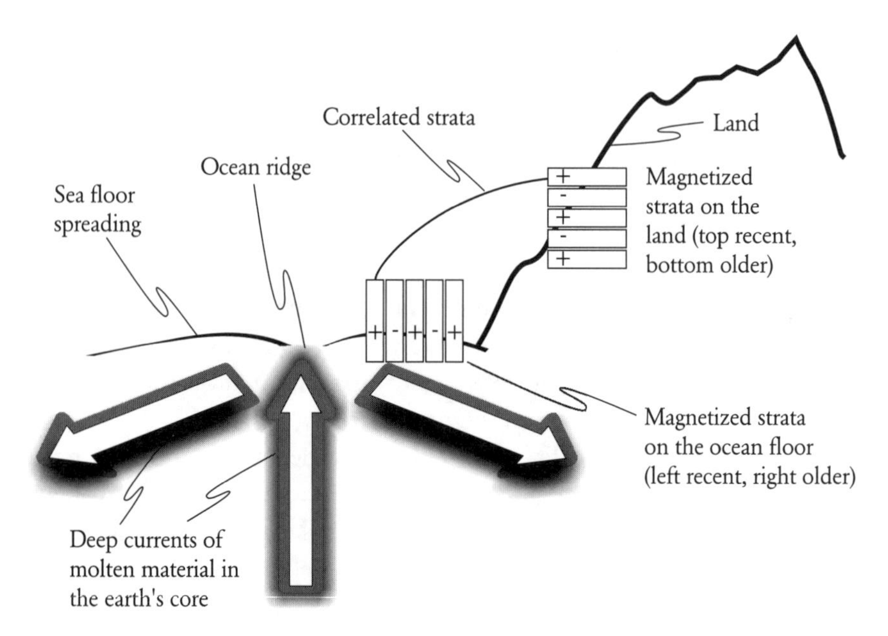

Figure 1 The magnetic records on the land and on the sea floor match, as predicted.

> suppose further, that prior to testing, *C* was thought to be unlikely (aside from its connection to *T*). Then *C* is strong evidence for the (approximate) correctness of *T.*

Here's a second example. In the early nineteenth century, Newtonian ideas were widely accepted, and Newton's corpuscular theory of light, in particular, was dominant. (This is the theory that light consists of tiny particles.) In one of the regular French Academy competitions, Augustin Fresnel submitted his wave theory of light. One of the judges, Poisson (a convinced Newtonian), examined the theory closely and deduced a surprising consequence, of which not even Fresnel was aware. The wave theory implied that a light source falling on a disk would cast a shadow on a screen behind. No surprise there. The surprising prediction was that there would be a bright spot in the middle of the shadow. (Crests and troughs of the waves cancel each other out leaving a shadow; but in the center a superposition of waves makes for a bright spot.) This unexpected consequence was thought to be absurd and grounds for rejecting Fresnel's theory as false. Nevertheless, an experiment was carefully performed. Amazingly, there was a bright spot, just as the theory predicted. Fresnel won the prize in 1818, and the wave theory of light grew rapidly in acceptance (see Figure 2).[3]

It's one thing to make up an "explanation" of any unexpected event after the fact, but to predict it in advance is either remarkable good luck or, more likely, a sign that the theory is on the right track. How, a typical working scientist asks, could any of this be a social construction?

Unification

Scientists make dramatic discoveries. Our newspapers often report amazing findings such as a black hole in our galaxy or perhaps a gene associated with cystic fibrosis. Working scientists are concerned with very much more than this. They're also concerned with the systematization of phenomena; that is, with bringing together diverse aspects of nature under the umbrella of a single theory. Newton, perhaps, provided the supreme example. Lucky Newton, thought Lagrange, there is only one universe and he got to explain it. So great was Newton's achievement that the Marquis de l'Hôpital seriously wondered if he ate and slept as other mortals do.

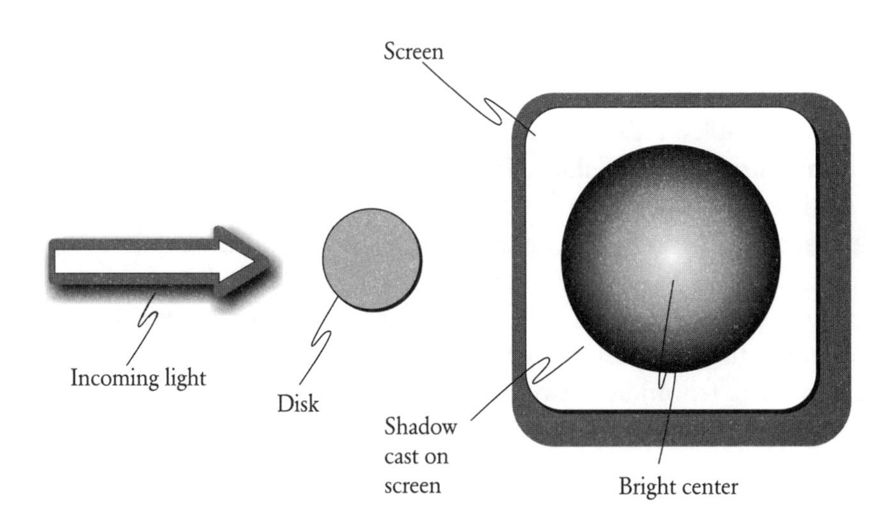

Figure 2 The predicted (though completely unexpected) bright spot in the middle of the shadow.

Isaac Newton's *Philosophiae Naturalis Principia Mathematica* (published in 1687) is arguably the greatest scientific work of all time. How did it come about? Edmond Halley (of comet fame) had wondered about the effects of an inverse square law; in particular, what would be the path of a planet orbiting the sun if the force between them is inversely proportional to the square of their distance apart? After discussions with Christopher Wren (the great architect) and Robert Hooke (another great seventeenth century scientist), he went to see Newton about it in Cambridge, England. It would move in an ellipse, said Newton, when asked. How did he know? Why, he had calculated it–but lost the paper. Newton promised to do it again and to send the result to Halley in London. With Halley's encouragement, however, Newton did so much more than that–he wrote the whole *Principia.*

Up to that time, a number of discoveries stood as rather isolated facts. Galileo, for instance, had discovered that bodies fell with an acceleration such that the distance fallen is proportional to the square of the time that has elapsed since the body was at rest. Kepler, on the other hand, had discovered a number of laws concerning planets, including the fact that they move in ellipses. Newton brought these and a great deal more under his theory. He started with three laws of motion: the law of inertia, the force law, and the action-reaction law. To the three laws of mechanics, which cover all physical interactions, he added the law of gravitation concerning one particular force.

This set of laws proved to be staggering in its power. From this single set of principles, Newton could explain the celestial laws of Kepler and the terrestrial laws of Galileo. The elliptic paths of planets and the rates of fall were derived from the three laws of mechanics and the law of gravitation. The falling of an arrow or an apple and the orbits of the moon or Mars were thus united as instances of a single general phenomenon, and so were a variety of other types of movement, such as the tides, the trajectories of cannonballs, and so on. A huge range of phenomena fell under Newton's theory. Much of it he made explicit. Even more became apparent in subsequent years when later generations of scientists continued to draw out the consequences. Euler, Laplace, Lagrange, Hamilton, and many others further elucidated the extent of Newton's profound achievement.

Reflecting on this example, we can extract another principle:

> *Unification.* Other things being equal, the theory most likely to be true is the one that explains the most facts and regularities using only a small number of concepts and first principles.

A second example of unification is worth a moment of our time. Darwin's theory of evolution was just as momentous a unification as was Newton's. Evolution was a popular idea in the nineteenth century; Darwin certainly wasn't the first to think of it. He did have a novel approach, however, one that relies on a mechanism he called "natural selection." In Darwin's theory there are a small number of basic principles: First, there is *variation* in any population; individuals differ in their characteristics. Second, there is *differential selection;* in a hostile environment where not every individual is going to endure, some characteristics are more likely to help the individual survive and reproduce than other characteristics (this is the so-called survival of the fittest). Third, there is *inheritance;* the biological characteristics of parents tend to be passed on to their offspring. Together these three principles imply the *evolutionary change of species.* This, in a nutshell, is Darwin's theory.

The glory of Darwin's account consists in the unification it brought to many distinct spheres of biology. In comparative anatomy, for example, one could explain why the arms of humans, the wings of birds, the legs of dogs, and the flippers of whales all have similar structures in spite of the fact that they have quite distinct uses–they come from a common distant ancestor. In biogeography one could explain the differences among species found in different locations. Why, for instance, would Darwin's famous finches on the Galapagos Islands be so different from island to island, and again so different from finches on the mainland? Being isolated, the evolutionary process specialized them for their particular locations. In paleontology (fossils), in embryology (development of organisms), in systematics (the distribution and grouping of organisms), in morphology (characteristics of organisms), and in taxonomy (classification) the explanatory power of Darwin's theory was evident. It unified diverse fields of biology like nothing before or since.

How, a typical scientist wonders, could a social constructivist account for this with any degree of plausibility?

Precision and Accuracy

Many theories account for trends: Because of a downturn in the economy unemployment rose last year; the presence of *el niño* means that winter this year is warmer than normal. These can be plausible explanations and useful bits of information, even though they aren't very precise. Occasionally a theory is stupendously accurate in its prediction. In this regard quantum electrodynamics (known as QED) has dazzled one and all.

Quantum electrodynamics is a theory built out of quantum mechanics and special relativity. As its name suggests, it applies to electrodynamic phenomena, the interaction of light and matter, and, in particular, spinning electrons. (Though it is misleading to do so, you can think of an electron as spinning like a little top. The trouble is, there is no way to imagine electron spin that is both intuitive and correct, so we tolerate this incorrect image–just be wary.) Since the electron is charged, a spinning electron is a charge in motion, a tiny current. This in turn will give rise to a magnetic field, and consequently, the electron has what is known as a magnetic moment. The stunning achievement of Julian Schwinger, Sin-itiro Tomonaga, and Richard Feynman (in 1948) was to show how QED could predict the value of this magnetic moment. At roughly the same time, but independently, Willis Lamb was able to make very precise measurements of its value. The degree of accuracy of the predicted value to the measured value was breathtaking–Nobel Prizes all around. Recent numbers will give you an idea:

measured value: 2×1.00115965221 ($\pm$ 0.00000000004);
predicted value: 2×1.00115965246 ($\pm$ 0.00000000020).

These numbers are in units of *magnetons*, $g = (e\hbar)/2mc$ (where e is the charge on the electron, $\hbar$ is Planck's constant divided by 2π, and c is the velocity of light). Writing about this years later, Feynman said, "To give you a feel for the accuracy of these numbers, it comes out something like this: If you were to measure the distance from Los Angeles to New York to this accuracy, it would be exact to the thickness of a human hair" (1985, 7).

If there is a methodological moral we can extract from this, it might be:

> *Precision.* Other things being equal, extreme accuracy in its predictions is a powerful reason for believing that the theory is correct.

Precision is a double-edged sword, as an example from the work of Kepler shows. Johannes Kepler (1571–1630) was a wonderful mathematician, very much in the God-is-a-great-geometer tradition of understanding nature. While still a student he became a convinced Copernican, but several remaining questions intrigued him. Why, for example, are there six planets (as was then believed)? Why are they at the distances they are from the sun? His brilliant guess linked the five Platonic solids with the orbits of the planets. At the center is the sun, with the sphere of Mercury around it. Now imagine the octahedron surrounding this sphere, then another sphere, this time of Venus, surrounding the octahedron. This is followed by the icosahedron, then the sphere of the earth. Next is the dodecahedron, then the sphere of Mars. After that the tetrahedron, the sphere of Jupiter, the cube, and finally the sphere of Saturn (see Figure 3).

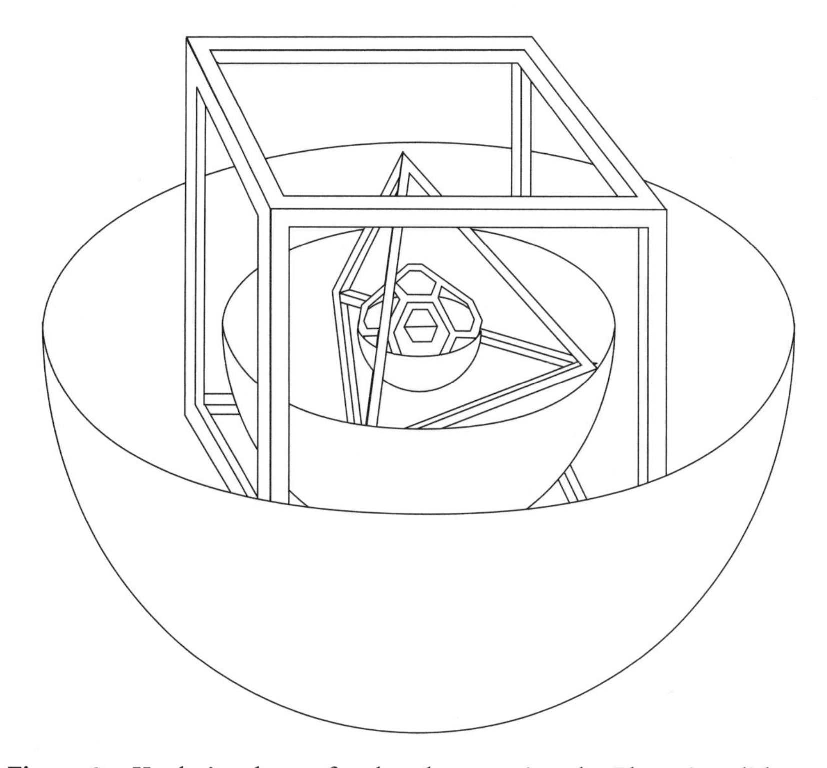

Figure 3 Kepler's scheme for the planets using the Platonic solids.

There are six planets with orbits at exactly the distances they are because the Creator made them that way; and God did it this way rather than some other, because, being a great geometer, he naturally loved the five regular polyhedra (the Platonic solids). It's hard to imagine a theory more aesthetically brilliant, or more pleasing to the mystical Kepler.

After this brilliant achievement, Kepler went to work for Tycho Brahe, the great astronomical observer. The data collected by Tycho over many years was unparalleled for its thoroughness, accuracy, and general reliability. After Tycho's death, Kepler became his successor and inherited the huge collection of astronomical data. He worked long and hard at trying to make his theory fit these data, especially the observations for Mars, but he could never get the fit better than 8′ of arc (i.e., an angle of 8 minutes; recall there are 60′ in one degree and 360 degrees in a full circle). By comparison, Copernicus had never expected better than 10′. To get a feel for how small the discrepancy is, a person with normal (unaided) vision can barely distinguish two stars that are separated by an angle of 4′. Kepler remarked,

> Since the divine goodness has given to us in Tycho Brahe a most careful observer, from whose observations the error of 8′ is shown in this calculation . . . it is right that we should with gratitude recognize and make use of this gift of God . . . For if I could have treated 8′ of longitude as negligible I should have already corrected sufficiently the hypothesis . . . But as they could not be neglected, these 8′ alone have led the way toward the complete reformation of astronomy, and have been made the subject-matter of a great part of this work. (Quoted in Barry 1898, 184)

The predictions and claims of Nostradamus are so vague we can declare them fulfilled no matter what happens. But how, scientists want to know, could social constructivism do full justice to the sort of accuracy that science often achieves?

Thought Experiments

Sometimes merely reflecting on the concepts involved can lead to the rejection of a false theory and point in the direction of a better one. Let's look at one of the finest examples of a thought experiment, young

Einstein's discovery of a tension between classical mechanics and electrodynamics.

According to Maxwell's theory of electrodynamics, light is an oscillation in the electromagnetic field. Maxwell's theory says that a *changing* electric field gives rise to a magnetic field, and a *changing* magnetic field gives rise to an electric field. If a charged particle is jiggled, it changes the surrounding electric field, which creates a magnetic field, which in turn creates an electric field, and so on. Maxwell's great discovery was that the wave created by this motion which travels through the electromagnetic field with velocity *c* is light.

When he was only sixteen Einstein wondered what it would be like to run so fast as to be able to catch up to the front of a beam of light. Perhaps it would be like running toward the shore from the end of a pier stretched out into the ocean with a wave coming in: there would be a hump in the water that is stationary with respect to the runner. However, it can't be like that since change is essential for a light wave; if either the electric or the magnetic field is static it will not give rise to the other and hence there will be no electromagnetic wave (see Figure 4).

> If I pursue a beam of light with the velocity *c* (velocity of light in a vacuum), I should observe such a beam of light as a spatially oscillatory electromagnetic field at rest. However, there seems to be no such thing, whether on the basis of experience or according to Maxwell's equations. (Einstein 1944, 53)

Conceptual considerations such as those brought on by this bit of youthful cleverness played a much greater role in the genesis of special relativity than worries about the Michelson-Morley experiment. Einstein goes on to describe the role of his thought experiment in later developments.

> From the very beginning it appeared to me intuitively clear that, judged from the standpoint of such an observer, everything would have to happen according to the same laws as for an observer who, relative to the earth, was at rest. For how, otherwise, should the first observer know, i.e., be able to determine, that he is in a state of fast uniform motion?

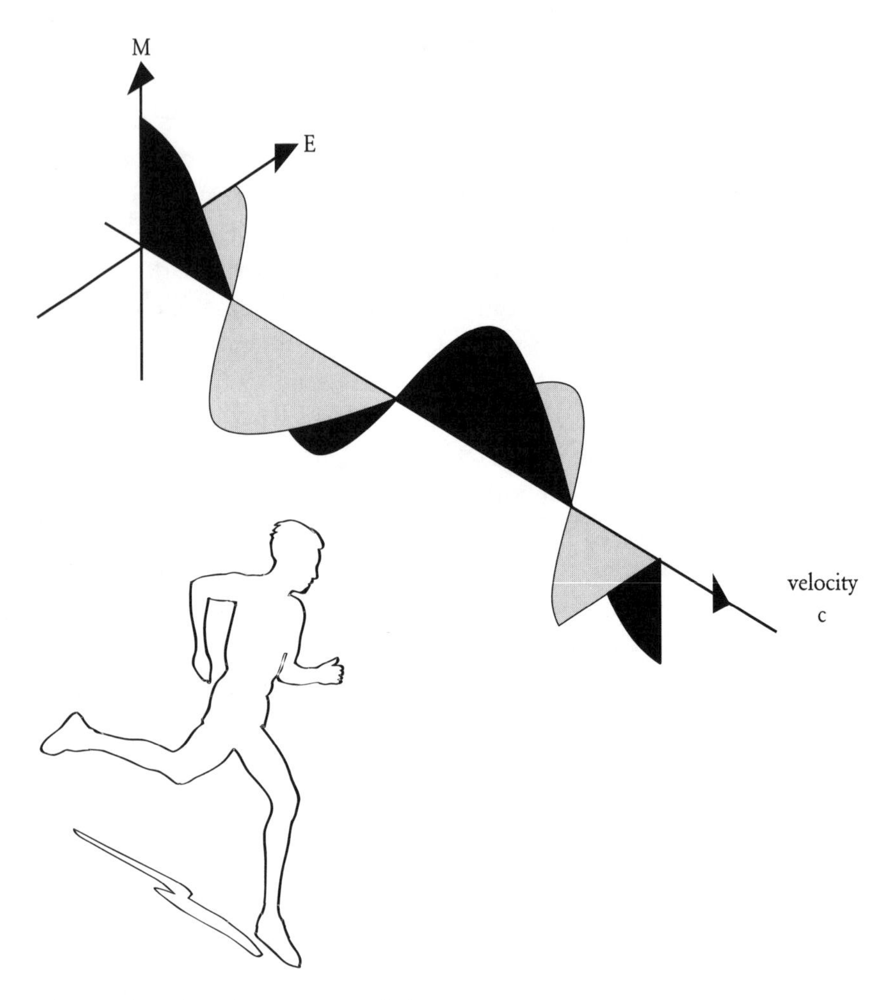

Figure 4 Einstein chasing a light beam.

> One sees that in this paradox the germ of the special relativity theory is already contained. (1944, 53)

And now to the moral, which should be obvious.

> *Thought experiments.* Conceptual clarification and even outright discovery can be achieved by the judicious use of thought experiments.

Here's another example to savor. Simon Stevin (1548–1620) produced a beautiful example concerning the inclined plane. Suppose we have a weight resting on a plane. It is easy to tell what will happen if the plane is vertical (the weight will freely fall) or if the plane is horizontal (the weight will remain at rest). But what will happen in the intermediate cases?

Stevin established a number of properties of the inclined plane; one of his greatest achievements was the result of an ingenious bit of reasoning. Consider a prism-like pair of inclined planes that are frictionless and that have linked weights such as a chain draped over it. How will the chain move?

There are three possibilities (see Figure 5): it will remain at rest; it will move to the left, perhaps because there is more mass on that side; it will move to the right, perhaps because the slope is steeper on that side. Stevin's answer is the first–it will remain in static equilibrium. The diagram on the right clearly indicates why. By adding the links at the bottom we make a closed loop that would rotate if the force on the left were not balanced by the force on the right. Thus, we would have made a perpetual motion machine, which is presumably impossible. The grand conclusion for mechanics drawn from this thought experiment is that when we have inclined planes of equal height then equal weights will act inversely proportional to the lengths of the planes.

The assumption of no perpetual motion machines is central to the argument, not only from a logical point of view, but perhaps psychologically as well. Ernst Mach, whose beautiful account of Stevin I have

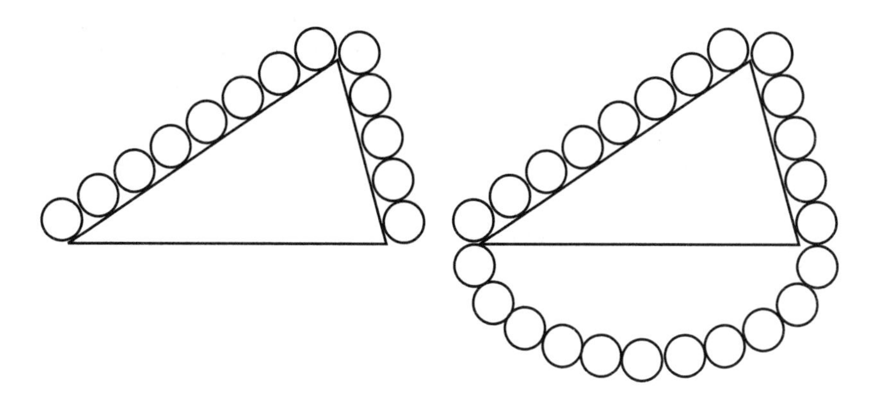

Figure 5 Stevin's thought experiment to show that the figure on the left is in static equilibrium.

followed, remarks: "Unquestionably in the assumption from which Stevin starts, that the endless chain does not move, there is contained primarily only a *purely instinctive* cognition. He feels at once, and we with him, that we have never observed anything like a motion of the kind referred to, that a thing of such a character does not exist" (1960, 34ff).

As we move step by step through a great thought experiment, it has a feel of inevitability about it, something that no made-up story has. Is there any hope, a typical scientist will ask, of social constructivists accounting for such wonderfully productive conceptual gymnastics?

Taking Refutations Seriously

Scientists love their children–their intellectual children–almost as much as the biological sort (and in some cases, more), but they take refutations seriously. The history of science is full of false starts: Ptolemaic astronomy, phlogiston chemistry, the caloric theory of heat, and the aether are a few of the more well known. Each of these was well developed and highly workable in its day. In the long run, however, each was found to be seriously wanting, so out it went.

We are often told that science is just another religion. No doubt there will be a number of striking similarities. Football is also called a religion, and again, there are lots of similarities. Of course, football is also called a science, and science sometimes called a sport. Are all these analogies helpful, or just tiresome fluff? The religion-science analogy fails to hold up at the most crucial point–commitment.

Every physicist allows that quantum mechanics could be deeply wrong. They all agree that tomorrow it may be overthrown by some quite unexpected experimental result or by some new and profound theoretical insight. Of course, it would take time to assess the new evidence; it would be rash to abandon such a wonderful theory too quickly. But in principle, it could go the way of earth-centered astronomy.

Could any Christian, by contrast, abandon belief in the divinity of Christ? Or abandon the belief that Christ died for our sins? Or abandon the belief that God is the creator of all things? The difference between a physicist and a priest is not so much in the subject matter.

Rather the difference is fundamentally this: a physicist can abandon all the central beliefs of current physics and still remain a physicist. A priest cannot abandon certain central beliefs without giving up the vocation. *Commitment* is a virtue in religion–and a sin in science.

Loyalty to our friends and family is an important social and psychological trait, valued and encouraged, but loyalty to a theory is despised. This is central to the scientist's self-image, and they wonder if any form of social constructivism can fairly account for this genuine feature of successful scientific activity that is apparently so at odds with one's natural inclination and self-interest.

Learning How to Learn

Scientific method isn't a completely eternal set of procedural verities. We have learned things about nature that have led to significant improvements in method.

Our first ideas about causal relations are based on simple instances. If I do *this, that* happens. So *this* causes *that.* The trouble is that often the causal situation is quite complicated; so instead of looking at specific examples, we set up large *study groups* and matching *control groups.* To check the efficaciousness of, say, a new drug, we select two groups of people randomly, giving the drug to only one group, then compare the results. Statistical techniques become crucially important: Are the two groups truly random? Are they representative samples? Learning how to investigate nature is an ongoing process. It was noticed that people's *beliefs and expectations* seem to have a bearing on their physical state. Those given the medicine thought it was going to be effective and their believing this contributed to their perception that they did feel some improvement in their condition. So the question arises, is it the drug or the belief about the drug that is having the causal impact? To check this, scientists keep the members of the study and control groups in the dark about who is getting the drug and who is getting a so-called placebo. This procedure is called a *blind test.* It didn't come naturally; scientists had to learn to do things this way.

They went on to learn even more. It was noticed that when researchers ask questions about how people feel after taking the drug, the questioners can inadvertently give cues as to what answer is expected, or worse, interpret a patient's remarks so as to fit what the researcher thinks or hopes is the correct experimental outcome. So the researchers who are directly examining the subjects are themselves kept in the dark about which group each subject is in. This is a so-called *double-blind test.*

Some aspects of scientific method–the use of the basic principles of logic, for example–may be eternal. Others are unquestionably tied to discoveries in the sciences themselves; blind and double-blind testing are due to advances in psychology. This is one of the reasons why no one could ever write a once and forever treatise on scientific method.

Let's make the principle explicit:

> *Methods improve.* Science is self-critical in the sense that discoveries in one area (psychology, optics) can lead to better ways of learning about how to investigate other fields (medicine, cell biology). The quality of the data is higher than when generated naively.

Leeuwenhoek was one of the first microscopists. He used a single lens and made some remarkable drawings of what he saw. Hooke invented the compound microscope, but there was little optimism about its ultimate utility, since, according to Newton, the problem of chromatic aberration could not be overcome. The problem is this: light of different colors has different focal lengths, so if blue light emitted from some object is in sharp focus, red light from the same source will be smeared. It turned out that Newton's problem could be overcome by a judicious arrangement of lenses made of different types of glass that have different refractive indices. In proper combination they could cancel out the aberration. One of the biggest advances came late in the nineteenth century when Ernst Abbe determined that the microscopic image is produced by the interference of light coming directly from the object and light that has been diffracted.

We certainly do not see microscopic entities in the same way that we see ordinary objects–but we do see them. Learning how the microscope works has greatly affected our beliefs about what we see. The theory of microscopy is much more than optics, however. It includes, for example, staining techniques. Much living material is transparent, so staining is essential to make it visible. All along there have been worries about the images viewed. To what extent are they "real" and to what extent are they artifacts of the viewing process? Stains often damage and kill cells. Is the structure of the viewed cell natural, or has it been seriously modified by the toxic stain? Are some of the objects seen really there, or are they produced by the optical properties of the lenses?

Beginning students of biology spend a significant amount of time learning some of the intricacies of the microscope. The same is true of astronomy students who learn about the peculiarities of the telescope. The naive observation of nature was long ago abandoned. We start off in a common-sense way, but we learn–sometimes quickly, sometimes painfully slowly–how to improve the way we learn. Not only are our theories subject to critical scrutiny, but our methods are similarly up for reevaluation.

There are many other aspects of the scientific experience, but this sample will do for now. The self-image of working scientists may be deeply flawed, but it is not cut from whole cloth. There are centuries of apparently spectacular achievement to which anyone can point. Regular working scientists cannot help asking whether social constructivists can even begin to do justice to this wealth of accomplishments. By sharing some of their experiences we begin to understand some of their outlook. Does it seem deluded?

The Anti-Constructivist Reaction

Working scientists as well as traditional historians and philosophers of science often react very negatively to social constructivist claims. Why do they have such a strong reaction? Sokal says that he doesn't fear what constructivists might do to science. His concern is political: the Left should be using science–not rejecting it–in the struggle for social justice.

Undoubtedly, this reaction is important and it accounts for much of the anger. There's clearly more at work than this, however. It's as if an appropriate sense of reverence were lacking. The historian I. B. Cohen asks, "Who, after studying Newton's magnificent contribution to thought, could deny that pure science exemplifies the creative accomplishments of the human spirit at its pinnacle?" (1985, 184). Speaking of the same achievement, another historian, Charles Gillispie, said, "it will always repay effort to study the mind and personality which founded science in generality and once for all. Fellow beings have a right to share in that triumph, and the duty to respect it. It enhances all humanity" (1960, 117). There is nothing new in idolizing Newton; the curious phrase is "the duty to respect" Newton's spectacular triumph–as if anything else would be like despoiling the Parthenon. For the admirers of science, social constructivists are not merely wrong–they are philistines and vandals scribbling graffiti on the temple of Athena.

3

How We Got to Where We Are

It's time for a short course in the philosophy of science. Many social constructivist doctrines are taken over directly from philosophers–often to the latter's chagrin. So seeing how such doctrines arise is important. As intellectual fields, philosophy and philosophy of science overlap significantly. A complete course in either would start with Thales, we would talk at length about (and with much affection for) Plato, then on to Aristotle, the medievals would get a look-in, Descartes would loom large, so would Locke and Leibniz; Kant, of course, is a major figure who couldn't be ignored–and still we haven't gotten beyond the eighteenth century. To keep this survey within fair compass, I will relate some of the highlights of the twentieth century only. Let's begin with the positivists.

Logical Positivism

The so-called logical positivists were a group who met regularly in the 1920s and early 1930s in the Austrian capital–hence the name "Vienna Circle." They included some of the most wonderful philosophers, scientists, and mathematicians of recent times: Moritz Schlick, Rudolph Carnap, Otto Neurath, Friedrich Weismann, Herbert Feigl, Phillip Frank, Karl Menger, Hans Hahn, and Kurt Gödel. They took Bertrand Russell's logic to heart (hence the "logical" in logical positivism), and seemed to be influenced by the early work of Wittgenstein. More important, however, was the philosophical influence of two giants. One was the spiritual father of the Circle, Ernst Mach, the great philosopher-scientist and fellow Viennese. The other was Einstein, who, by the way, also acknowledged a great intellectual debt to Mach, the staunch champion of empiricism and sworn enemy of all metaphysics. Mach put his empiricism to work criticizing Newton's absolute space (it can't be experienced, so into the trash it goes) and espousing what is now known as Mach's principle (only rotations relative to the observable stars, rather than to so-called absolute space, are empirically acceptable; this is the origin of inertia). Einstein, with his dramatic use of rods, clocks, and observers, and his rejection (in the theory of relativity) of such unobservable notions as absolute simultaneity, seemed to have found the methodological key to investigating the world.

These were the positivists' heroes. They developed the doctrines of Russell, Mach, and Einstein and they spread the word with missionary zeal. Positivist attitudes have permeated the sciences, even when there has been no direct influence from the Vienna Circle. American "pragmatism" developed by Peirce, James, and Dewey, and Percy Bridgman's "operationalism" are kindred doctrines.

Members of no large collection of people can be expected to have identical philosophical beliefs, but there was a surprising degree of common outlook, an outlook that positivists liked to call scientific philosophy. They even published a manifesto: *The Scientific Conception of the World: The Vienna Circle.*[1] Positivist principles or instincts include two central, related items: *empiricism* (all knowledge is based on sensory experience) and *verificationism* (to be meaningful, a statement must be empirically testable; often the claim was that the meaning *is* the method of verification). Spelling out some details of these two notions will be useful.

There has been a constant battle throughout the history of philosophy. Empiricists claim that all knowledge is based on sensory experience. Concepts such as "red" and "yellow" or "apple" and "banana" can only be acquired by seeing apples or bananas, by perceiving red or yellow things. The truth or falsity of propositions such as "Bananas are yellow" and "Apples are purple" can only be tested by sensory experience. Empiricism's historical rival is rationalism, the doctrine that claims that *some* of our knowledge has a nonsensory source. Perhaps some of the things we know are innate, or perhaps we can "see" some truths with the "mind's eye." Rationalism has always had the upper hand when it comes to mathematics–perfect circles and imaginary numbers are not the stuff of sensory experience. Ethics, too, is problematic for an empiricist–no empirical observation tells us that murder is wrong.

Having introduced the term "rationalism," I must add a caution. One sense of the term contrasts with empiricism; they are rival accounts of how we know. In the science wars, however, the term comes up repeatedly in contrast with "social constructivism." In the latter context rationalism just means that scientific decisions are based on reason and evidence. In this sense even staunch empiricists (who deny any form of nonsensory knowledge) are rationalists. Context should make it clear which sense is intended, but aside from this chapter it is the more encompassing sense of rationalism that will almost always be at work.

Common sense seems on the side of empiricism and against the sort of nonsensory sources of knowledge that a traditional rationalist would uphold. Everyday examples fit the bill perfectly. "The coffee is cold," "There's still some pizza left," "The dog has tracked in a lot of mud" are all straightforward examples of empirical statements. The big problem is this: Can empiricism do justice to all of science? No one can actually see electrons, magnetic fields, genes, entropy, or Freudian superegos. What sense can we make of these concepts?

The positivist approach to this problem was to split the language of science into two distinct vocabularies, based on the apparent fact that some of the things we talk about we can see and some we can't. *Observational terms* would include such expressions as: "apple," "red," "hot," "white streak in cloud chamber," and so on, whereas *theoretical terms* include, among others: "positron," "DNA," "quasar," "id," and "superego." We know the meaning of an observational term when we acquire the relevant experience. I know what "apple" means because I have seen (and touched and tasted) them. The truth of statements that contain only observation terms can usually be tested by direct sensory experience. We need only *look* in order to determine that strawberries are red or that there is a white streak in the cloud chamber. The difficulty is to give some sort of meaning to theoretical terms, and the only way to do that is to link them in some appropriate way to observation terms. Here's a slightly artificial example [*with running commentary italicized and in brackets*] to see how it works. It has the form of a theory explaining (or predicting) the existence of a white streak in a cloud chamber.

1. (Theory): Electrons are particles of mass *m* and charge *e*. [*The terms "electron," "mass" and "charge" are theoretical; we have no idea what they mean, at least not at this stage. We could take this statement as a definition of "electron," but since "mass" and "charge" are also meaningless at this point, so is "electron."*]
2. (Theory): An electron has moved through the cloud chamber. [*This statement has the form of an initial condition; it's not a claim about electrons in general, but an assertion about a particular electron. The term "cloud chamber" is interpreted in experience (since we can see the cloud chamber in front of us), but at this stage "electron" is still a meaningless word.*]
3. (Theory): A charged particle moving near a molecule will ionize the molecule. [*This statement has the form of a law of nature. The terms "charged," "molecule," and "ionize" are all theoretical, hence the statement is at this stage without meaning.*]
4. (Theory): A series of ionized molecules in a cloud chamber will scatter electromagnetic radiation at frequency *f*. [*This statement also has the form of a law of nature. The terms "ionized molecule" and "electromagnetic radiation" are theoretical, hence the statement is at this stage without meaning.*]
5. (Bridge principle): A series of ionized molecules that scatter electromagnetic radiation at frequency *f* is a white streak. [*For the positivists, a so-called bridge principle is a definition; it stipulates the meaning of a theoretical term, "series of ionized molecules that scatter electromagnetic radiation at frequency* f,*" by means of an observation term, "white streak." It is crucial to note that this is not a claim about nature; rather it stipulates how to use the language.*]
6. (Observation): There is a white streak in the cloud chamber. [*This statement uses observation terms only, so it is completely meaningful and its truth-value can be directly checked by sensory experience.*]

The first four statements above constitute the theory. Taken together they explain or predict the final statement; and the final statement, being directly testable through observation, is in turn used to test the theory.

One way of understanding all of this is to say that theoretical terms pick up their meaning *indirectly* by being linked via bridge principles to observation terms. Another way is to simply say that theoretical terms are *not* given any meaning at all; they are simply very useful calculating devices. In any case, this is how positivists made sense of entities that couldn't be experienced. Their solution to the problem of theoretical entities is largely linguistic–we have no idea what electrons are, but we do know how to deal very successfully with the word "electron." That is sufficient for what really matters to any empiricist, namely, organizing and predicting our sensory experiences.

A focus on language is central to positivism, and the theory of meaning played a central role in their thinking. The commonly accepted positivist view was *verificationism,* which tied meaning to method of testing. No formulation was ever agreed upon, but the underlying idea is easy enough to grasp. In order for a statement to be meaningful (that is, to be true or to be false), there must be an empirical method that would determine the truth or falsity of the statement. Clearly, "Apples are red" and "Bananas are blue" both pass the test for meaningfulness, for in each case we can tell whether the statement is true or false by simple inspection.

What about "Protons are more massive than electrons"? Since both are charged particles, they will be deflected when moving at the same velocity through a magnetic field. In a cloud chamber this will show up as white streaks of different radii of curvature–the more massive the particle, the straighter its path. The truth of this statement would have to be spelled out using bridge principles and so on. Though complicated, a technique is nevertheless available, so the claim "Protons are more massive than electrons" is a perfectly meaningful claim, since there is an empirically acceptable method of checking its truth.

The verification principle came in degrees of strictness. The most stringent demanded that a statement be actually testable. A less strict version would allow "There is a planet with intelligent life in the galaxy Andromeda" to be meaningful because we could at least describe a method–send a rocket to check the planetary systems of each star in Andromeda–even though carrying it out would not be practically possible. This means that the statement would be meaningful–true or false–though we may never know which. With a sufficiently liberal version of verificationism even "There is an afterlife in which God rewards the faithful and punishes the wicked" becomes meaningful. How could you test it? Well, wait until you die, then . . .

Though examples of the last sort were proposed, most self-respecting positivists considered religious sentiments as paradigm instances of meaningless statements. Religious beliefs, most of Hegel's writings, and claims such as "The Aryan Race is superior to others" were taken to be nonsense (i.e., not meaningful). This last example is particularly noteworthy. We can imagine what the political climate of Vienna was like in the late 1920s and early 1930s. Fascism was on the rise and a genuine threat to civilized life. The members of the Vienna Circle were decidedly on the Left, mainly Social Democrats and Marxists. Many were also Jews. All were deeply opposed to Hitler. The same, alas, cannot be said of Heidegger and many other German and Austrian philosophers who were unsympathetic to science. One of the travesties of current science studies debates is the branding of the positivists as political reactionaries. Such ignorance is culpable. Dealing with political issues meant as much or more to Neurath, Carnap, Frank, and Hahn as coming to grips with science for its own sake.

The role of theoretical entities in the positivist picture is one of helping to organize and systematize observable experience. This is quite different from theorizing in the hope of understanding what is *really* going on–the positivists were not terribly keen on looking for the hidden causes of things. Indeed, in their view it is generally meaningless to talk about such things. Nevertheless, there was a rather elegant theory of explanation (due to Hempel) that went along with the rest of the positivist outlook. An explanation is an *argument* in which the theory and the initial conditions serve as premises. The thing we want explained we deduce as a conclusion from those premises. Why is there an eclipse of the sun at time t? To explain this we state the theory T, which, let us suppose, describes the motion of the planets around the sun and the moon around the earth. We often need to assume an auxiliary theory, AT; in this case it's the assumption that light comes from the sun in straight lines to the earth and that this light could be blocked by a material body such as the moon placed between us and the sun. We also state the initial conditions, IC, the exact positions and velocities of the earth, moon, and sun at some earlier time t'. From these premises we then derive the conclusion, *E:* There is an eclipse of the sun at time *t*. The form is simple:

T (theory describing general motion of earth, moon, and sun)
IC (initial condition stating position and velocity of moon, sun, and earth at time *t′*)
AT (auxiliary theory, e.g., optics)
∴ *E* (statement that eclipse occurs at time *t*)

This has the general form of an explanation. It involves a wonderful symmetry with prediction. If we run through the argument *after* the eclipse has occurred we are *explaining* it, but if we do the derivation *before,* we are *predicting* the eclipse. Anyone with a soul can feel the elegance of this symmetry, and since prediction is tied to confirmation (making correct predictions is a sign that the theory is on the right track), there is a grand linkage of three fundamental concepts: explanation, prediction, and confirmation.

Naturally, all of this was too good to be true. The verification principle was a powerful tool for cutting through nonsense–too powerful. Although much of science came off looking good and Hitlerian doctrines were found to be nonsense, several branches of knowledge were in danger of being lost. Ethics, mathematics, and even philosophy itself were all problematic. None of them was open to empirical testing, so each seemed to lack cognitive significance. It was all well and good to be rid of Hitler and perhaps of Hegel–but arithmetic?

The various responses to this predicament were often interesting. One popular account of ethics, for example, was *emotivism.* A moral pronouncement such as "Murder is wrong" has no truth-value on this account. Instead it is an expression of preference or sentiment; in particular it expresses a strong disapproval of killing people under certain circumstances. Similarly, when I say "Sunsets are beautiful" I should be taken as expressing a liking for sunsets and not as saying something objectively true about them. Likes and dislikes can be tested, of course, but there is no objective matter of fact about ethical or aesthetic right and wrong.

The positivist solution to the mathematical problem was–not surprisingly–to be found in language. Some of our statements report matters of fact; for example, "John is a bachelor." Others tell us nothing about the world, but instead report the use of words: "All bachelors are unmarried men." You can determine that the second statement is true by looking in a dictionary, since its truth is wholly dependent on the meaning of the terms involved. On the other hand, to determine that John is a bachelor will require genuine empirical inquiry. All of mathematics, according to a common positivist account, is based on the meaning of the terms involved. "5+7=12" is a true statement because of the meanings of "5," "+," "7," "=," and "12," not because there is an independent realm of mathematical facts that mathematicians are correctly describing.

What about philosophy itself? Critics were quick to point out that the verification principle itself is not verifiable. Does this mean that it, and indeed all of philosophy, is meaningless? Yes and no. In the strict sense, yes, philosophy does lack cognitive content, according to the positivists. It is quite unlike, say, physics or geology. But not all unverifiable propositions are gibberish. Some propositions–and the verification principle is, on their view, a perfect example–can offer us a useful guide to life. Wittgenstein, in his *Tractatus,* remarked that anyone who understood his book would realize that it must be strictly nonsense. Nevertheless, he added, one uses the book as a ladder to climb to where things are seen properly. Once there, the ladder can be thrown away, for now we see the world aright. Similarly, the positivists might say, though verificationism cannot itself be verified, adopting it leads to an intellectually healthy outlook.

Let me quickly summarize in point form some of the chief principles of positivism (*italicized*) and their problems.

- *There is a sharp distinction between theory and observation; the latter is neutral, independent of any theory or background beliefs.* This has been one of the most contentious of empiricist doctrines. Thanks to Kuhn and others (as we shall see below), this point is largely discredited today.
- *Nothing is meaningful unless it is empirically testable. Consequently, nothing is meaningful (in the strict sense) except empirical science.* Variations on this principle have always been around and likely always will. The instinct of some (known as realists) is to sharply separate truth from what can be known, and to acknowledge (perhaps even with a touch of humility) that our puny efforts can't hope to grasp all of reality.

Verificationists, by contrast, have the opposite instinct. They often mock realists for their arrogant assumption that there could be a "God's-eye-view," and claim that the only truth worth talking about is humanly accessible truth. The instincts that lie behind the idea of a "God's-eye-view"–both pro and con–play a big role in the science wars.

- *History of science is cumulative. There are no revolutions in genuine science. Since science is chiefly concerned with the observable part, the history of science is largely a history of increasing our empirical knowledge. Later theories don't really overthrow earlier theories; instead they go beyond and leave the earlier theory as a special case.* Both Popper and Kuhn (as we will shortly see), completely reject this–and rightly so. The history of science is revolutionary; old theories are refuted and replaced by new ones.
- *Science organizes our experience and predicts what we will observe, but does not try to give deep causal explanations.* This remains a contentious point today. Although many who share positivist instincts think it quite correct, the "realist" opponents of positivism often glory in the boldness of science to go beyond appearances and attempt to grasp reality.
- *Science aims at empirically adequate theories, not "true" ones. Since the theoretical terms of science are only indirectly defined via observation terms (or possibly not defined at all), there is little point in claiming that they "hook onto" the world.* This sort of anti-realism has a very long and distinguished history. In astronomy, for example, since the time of Plato it has often been maintained that we have no access to the heavens; so it is hopeless to get a true description of them. All we can do is "save the phenomena." We can make up stories about how the heavens work; the only thing that matters is that a theory correctly predict where and when various points of light will be. Nothing else is possible, so it's folly to even try. In recent years much ink has been spilled on this question, without decisive results.
- *There is a distinction between discovery (thinking up new ideas) and justification (objective testing of new ideas). When it comes to testing, it matters not, for instance, what a person's politics are; all that matters is that her theory is objectively verified. Any bias she might have in creating her theory is filtered out in the testing process.* This remains a very contentious point

with a wide spectrum of opinion, but no consensus. It's crucial in the science wars, for if discovery and justification are intimately linked, then any bias present in the creation of the theory might not be eliminated in testing.

I'll explain the topic of *realism vs. anti-realism* and the *objective vs. subjective* issue in more detail in the next chapter.

Popper

Karl Popper was the chief rival of the positivists, especially during the middle years of the twentieth century. He has been quite influential, though not always in ways he would have wished or approved.

Popper made a career out of a simple logical point. Theories often have the form: *All* A*s are* B*s*. (All ravens are black. All chemical processes are mediated by phlogiston. All bodies attract one another with a force proportional to the product of their masses and inversely proportional to the square of their distance apart.) To refute such a theory one need only find a single *A* that is not a *B*. However, to confirm such a theory, says Popper, one would have to check every single *A* to ensure that it is also a *B*. This would require an exhaustive search of the entire universe–past, present, and future. From a logical point of view decisive refutations are possible, but confirmations are not.

This little point has a big moral, says Popper: The proper method of science consists of *conjectures and refutations*. It is hopeless to try to confirm any theory, but we can make scientific progress through trial and error: make a bold guess, then try to refute it.

Popper was quite interested in the boundary between science and nonscience–the demarcation problem, as he called it. His solution is simple: To be scientific is to be refutable (in principle). That is, the difference between a genuine scientific theory and any piece of pseudoscience lies in its openness to refutation. An astronomical theory that predicts an eclipse next Tuesday afternoon is sticking its neck out. If there is no eclipse, the theory is refuted. It is this *possibility of refutation* that makes it a real science. By contrast, a theory that can explain anything, come what may, is not to be taken seriously.

Popper had his favorite *bêtes noires,* Marx and Freud. The problem with Marxism, as he saw it, is that Marxists adapt too easily to anything that happens. Everyday events are interpreted in such a way as to confirm their general outlook. If predictions of social unrest fail to materialize, then the theory (in the hands of its advocates) seems to have no trouble explaining the difficulty away. Freudian psychoanalysis is the same, or even worse. Popper delighted in the retelling of one particular story. Freud's theory said a particular patient had a particular condition, and that condition ruled out having a particular kind of dream. Amazingly, the patient had that very dream. According to Popper, the Freudian analyst should have taken this as a refutation of the theory, but instead the analyst declared that the patient had transferred his hostility from his father onto the analyst. The best way to hurt the analyst would be to "refute" his theory. That, according to the analyst, is why the patient had the otherwise unexpected dream. Victory snatched from the jaws of refutation! Of course, it smacks of dishonesty. Those who seek confirmations rather than refutations lack the critical spirit, says Popper, sometimes to the point of being intellectual frauds.

Falsifiability demarcates science from nonscience, but it is quite unlike the verification principle, which provides the mark of the meaningful. Popper took lots of nonfalsifiable things to be perfectly meaningful–ethics, metaphysics, even branches of science that failed the falsifiability test. Darwin's theory of evolution (at least for the most part) fails to make any testable predictions. It provides wonderful explanations for how things came to be as they are, but it offers no novel predictions about the future. (It does, however, make so-called retrodictions, predictions about the past. The existence of archeopteryx–a species with characteristics of dinosaurs and of birds–could be seen as a successful retrodiction.) The problem of predictions in evolutionary theory should not be surprising. We would need to grasp an unmanageably large amount of information in order to calculate what would happen next. Remember, we can hardly predict next week's weather, so how could we possibly predict, say, *homo sapiens*'s development over the next million years? Popper happily accepted what he called "metaphysical research programmes," which is how he classified Darwin's theory.

The falsifiability principle distinguished science from nonscience, but it couldn't tell the difference between intellectually good stuff and rubbish. At this point all Popper could do is appeal to a general principle–the critical spirit. Darwinians have it; religious fundamentalists who insist on the account of creation in the book of Genesis lack it. But this is much too vague to be of any use, and fundamentalists recognized the opportunity they had been handed on a platter. Here was Popper, a famous philosopher of science, saying that Darwin's theory is not real science. So-called creation scientists claimed that they were at least no less scientific than that, so they should be given equal time in the public schools to teach "creation science." Needless to say, Popper, a professed lifelong socialist and atheist, was horrified at the use some right-wing Christians were making of his cherished falsifiability.

Popper was an empiricist of sorts, but in detail differed considerably from the positivists. Concepts are taken *to* experience, not (*contra* the positivists) derived from experience. There is no significant observational/theoretical distinction for Popper: protons and pine trees are not different in kind. Science makes conjectures about both and those conjectures are equally open to refutation in principle. Popper was also a common-sense realist; that is, he took scientific theories to be claims about how things really are, not just useful instruments for organizing and predicting experience.

Naturally, there are numerous fine points that I must pass over here for lack of space. A fuller account would mention at length that theories are tested by their singular spacetime consequences, e.g., "The swan at place *p,* at time *t,* is white." This is a *basic statement,* as Popper called it. Basic statements are tentatively accepted by convention. They cannot be accepted because of experience, since they are just as theoretical as any other. For Popper, observations are loaded with theory.

When a scientific theory is put to the test and fails, it is, of course, *refuted.* When it passes, however, the theory is said to be *corroborated.* Popper is adamant that this is not the same as *confirmed,* but the notion has baffled his readers. If "corroborated" just means "wasn't refuted this time," then the notion is innocuous. Yet Popper seems to want more than that. Commentators have found it hard to see anything here but a notion of inductive support, however abhorrent that may be to Popper.

The general idea behind Popper's approach should be clear: Science is a bold enterprise, in which heroic scientists make daring claims about nature; but they are typically intellectually honest and thus willing to take defeat when it comes, as it almost inevitably does. Popper's picture is completely contrary to that image of the cautious, careful scientist who diligently collects data as a squirrel collects nuts and who then guardedly makes only the most circumspect inferences. Science, says Popper, is revolutionary. It is not the cumulative acquisition of carefully established facts. It is also more than the organization of observable experience; science is the bold and unending search for a deep explanation and understanding of reality.

Popper's image is almost romantic. It is no surprise that so many working scientists are pleased with this picture of themselves. Admirers of Popper include Herman Bondi, of steady-state cosmology fame, Macfarlane Burnet who won a Nobel prize for work in immunology, John Eccles, a Nobel prize-winner for work in neurophysiology, and Peter Medawar, also a Nobel prize-winner for work in immunology, all pillars of the British scientific establishment, and like Sir Karl himself, all knights of the realm. So why hasn't Popper's account come to be dominant? I've mentioned some of the issues; I'll now summarize Popper's views (*italicized*) and mention a few more difficulties.

- *Science is a process of bold conjectures and attempted refutations.* This seems to ignore the fact that there are lots of scientific activities that don't fit this pattern at all; some activities are not attempts to explain, but are rather a search for interesting new phenomena. It also ignores the fact (correctly stressed by Kuhn) that much of science is puzzle-solving, an attempt to make things fit into an already accepted pattern of explanation.
- *There is no logic of discovery, no method of having good ideas.* The only place that rationality can play a role is in the testing of theories after they have been proposed. It's no small irony that Popper's most famous book, *The Logic of Scientific Discovery,* denies the very existence of the subject matter of its title. The distinction between discovery and justification, however, is a point on which Popper and his positivist rivals completely agree. It's a very important issue from the point of view of the science wars, since it claims, in effect, that a person's politics and other social interests are completely irrelevant in the process of theory-testing, because all bias can be filtered out in the process of justification. Kuhn and many others rightly deny this distinction.
- *A single counterexample refutes a theory.* Theories do not by themselves imply observations. They usually do this only in conjunction with as-

sumptions about initial conditions and auxiliary theories. If the conjunction of these premises implies something false, then clearly, at least one of the assumptions is false. But which one? We could blame the theory, but the fault may lie with the auxiliary theory, or perhaps we got the initial conditions wrong. This is often called the Duhem (or Duhem-Quine) problem. Popper makes refutation look easy. It is anything but. The problem has been cited repeatedly by social constructivists to show that rational theory-choice is an illusion. This is closely related to the infamous problem of *the underdetermination of theories by evidence,* a problem I'll take up at length in a later chapter.

- *Theories are scientific insofar as they are refutable. Any theory that cannot in principle be refuted is simply not part of science.* This rules out as nonscience such bold claims as: "There is life elsewhere in the universe." Since this claim does not say where or when this life exists, it would take an exhaustive search of the entire universe, past, present, and future, to refute such a claim. Since this is not possible, the claim is not refutable, and so it is not scientific, according to Popper. As I mentioned earlier, some wonderful explanatory theories such as Darwinian evolution fail to make the grade. So much the worse for Popper.
- *Any form of inductive confirmation is hopeless. Positive evidence in support of a theory is an illusion.* This seems far-fetched, to say the least. Making sense of induction is a central part of coming to understand what scientific method really is. But let's be as charitable to Popper as we can. Perhaps the Popperian attitude is plausible when applied to general relativity, or quantum chromodynamics, or some other piece of speculative, pure science that plays no role in practical affairs or daily life. But what about our faith in technology? When I get on an airplane, am I really making a bold conjecture that it can fly? I'm much too cowardly for that. Popper's position demands a reckless attitude to daily life or a hopeless schizophrenia–induction is present in practical matters but absent in pure science.
- *The scientific spirit is the critical spirit.* To an important extent this is certainly true. But the history of science seems at odds with this, at least on occasion. Elements of dogma have been present and seem even to have been useful. Attempts to make things fit rather than attempts to overthrow have often prevailed in the history of science, including the history of the best science. We'll see more of this

when we look at Kuhn momentarily. It would be wrong to overstress Popper's shortcomings on this point, however, for it really needs only mild qualification. Popper's enthusiasm for the critical spirit is unquestionably an enduring contribution to the betterment of all of intellectual life.

KUHN

Thomas Kuhn's *Structure of Scientific Revolutions* (1962, 2nd ed. 1970) aims to show how science is really done. It is an attempted description of actual scientific practice. Philosophers, on the other hand, are largely concerned with how science *ought* to be done. However, the gulf between the *is* and the *ought* had better not be too large. If real working scientists are all irrational by the lights of some normative philosophy of science, then this might well be seen as a *reductio ad absurdum* of the philosophical theory. This is why Kuhn's book is so important to both historians and philosophers of science.

Kuhn has a couple of axes to grind. He disagrees with the positivist position that the history of science has been cumulative; there have been a number of very significant revolutions that overturned the most basic beliefs of earlier scientists. He also disagrees with Popper's view that scientific history has been filled with incessant criticism; science often proceeds in an inductive, cumulative way that leaves foundational issues utterly unquestioned.

An example of cumulativeness is the (alleged) case of Newton's gravitation theory superseding and explaining Kepler's laws. Kepler's three laws say that (1) planets move in ellipses with the sun at one focus, (2) equal areas are swept out in equal times, and (3) the cube of the radius divided by the square of the period is a constant for all planets.

Newton's laws, it is often claimed, imply Kepler's.[2] Thus Kepler's theory is reducible to Newton's. Science marches on, abandoning nothing; later theories are merely more general and contain their predecessors. Thus, we have a kind of Russian doll model (smaller dolls contained in larger dolls that are contained in yet larger dolls) of the history of science. This is what many positivists thought the actual history of science looked like. It is also the picture given in many science textbooks and the picture held by many working scientists of their own fields. Past science, say the holders of this view, was correct as far as it went and later science just goes further.

For Popper, once again, the history of science is a history of bold conjectures that are submitted to crucial tests. Any theory that does not pass a test is tossed out. The history of science is not cumulative, on Popper's view, but revolutionary; it is littered with refuted theories. Kuhn finds both of these pictures false: Science is not cumulative; it is revolutionary. But it is not constantly revolutionary; there are periods when criticism stops. Let's take a look at some of the details of Kuhn's model of scientific change.

- *Normal science:* Most science is normal science. It is what is done by scientists who agree on all the basics. This means that they agree on what the world is made of, on how things interact with one another, on symbolic representations (everyone working in the Newtonian paradigm accepts the expression "$F = ma$," for instance), on what the outstanding problems are, and on the correct way to tackle them. And they all agree on whether a proposed solution really does or does not work. Basics are taken for granted; the fundamental theoretical framework is beyond criticism.
- *Paradigms:* Normal science is a puzzle-solving activity. It is an attempt to fit things into a pattern. The pattern is set by a paradigm. This is a concrete example, a specific bit of scientific work that all take to be a great achievement. A paradigm is like a legal precedent. A paradigm/precedent has no explicit directives associated with it, but close study suggests how to treat similar cases. In learning an inflected language such as Latin, we are given paradigms. Thus the first declension Latin noun *porta* (gate) is declined:

	Singular	Plural
Nominative:	*porta*	*portae*
Genitive:	*portae*	*portarum*
Dative:	*portae*	*portis*
Accusative:	*portam*	*portas*
Ablative:	*porta*	*portis*

When encountering other first declension nouns (e.g., *fama, fortuna, philosophia*) we know from the paradigm how to decline them.

Explicit rules play little or no role in guiding us. The example, the paradigm, is everything.

A paradigm generates certain associative practices, including basic ontology (what the world is made of), assumptions as to which are the important problems, heuristic norms (how to go about doing normal science), and evaluative norms (criteria for judging proposed solutions to problems).

To hold a paradigm is to adopt a Wittgensteinian "form of life," a *weltanschauung.* During the reign of a paradigm the course of science is cumulative.

- *Crisis:* Some of the puzzles and anomalies of normal science are very stubborn. When several persist for a long time a period of crisis sets in. The admission of crisis only comes with the greatest reluctance. ("It is a poor worker who blames his own tools.") Crisis induces attacks on the paradigm itself; basic assumptions are questioned. Many begin to consider anomalies as not just tough problems but rather as outright counterexamples to the theory.
- *Extraordinary science:* Normal science is no more; the united community has disintegrated into several diverse schools. All criticize all others. Several new paradigms come into existence. All kinds of devices are used to persuade others to convert–even philosophical arguments, says Kuhn. Eventually one of the rival paradigms comes to prevail. The vast majority of scientists are won over to the new theory. This is a scientific revolution. The conversion that takes place is much like a gestalt shift (see Figure 6, the duck-rabbit example below).
- *Entrenchment of the revolution:* The way a revolution in science becomes entrenched is not by further empirical evidence or argument, but rather by a kind of indoctrination. Students learn the new theory from textbooks, which present the exemplars that *illustrate* the theory as if they were *independent evidence* for it. Textbooks often give a short history of the subject that makes the paradigm look reasonable and old rivals look silly. (Just consider how Galileo's rivals are often depicted today–as dogmatists who refused to look through his telescope.) It is the winners who write the books. Kuhn is not critical of this dogmatic approach; he thinks it helpful for practicing scientists not to be skeptical about basics. Students learn by example in

the lab; and much of what they learn is *knowing how* as well as *knowing that.* This nonpropositional knowledge plays a big role in how one does science. (You know how to ride a bike, but try to articulate it!) Kuhn cites Max Planck (the great German scientist who founded quantum theory): "A new scientific truth does not triumph by convincing its opponents and making them see the light, but rather because its opponents eventually die, and a new generation grows up that is familiar with it." Thus, we return to (a new) normal science.

So much for our quick look at Kuhn's view of science and scientific change. Several interesting and contentious points arise from his account.

- *Theory-ladenness of observation:* The thesis, roughly, is that seeing is *seeing as* or *seeing that,* and never neutral, passive observation. Our beliefs and expectations condition what we see. Perception is paradigm dependent. There is no "given" that is independent of all conceptualization, no "neutral observation" that could be the independent empirical basis for testing rival theories. Under certain conditions, if we expect to see a duck, we will; but if we expect to see a rabbit, we will see that instead.
- *Meaning and incommensurability:* The meaning of a term is dependent on the role it plays in a theory. It is quite remarkable how much Kuhn took over from the positivists. Though he rejected their distinction between theory and observation, he did accept the related positivist view of language. In this view, key terms are defined contextually; that is, they get their meaning from the role they play in the theory. Consequently a change of theory involves a change in meaning. Even though the term "energy" may appear in several theories, it nevertheless means something quite distinct each time. Scientists with different paradigms may seem to say contradictory things about energy, but actually they are talking about different things. The different theories, says Kuhn, are incommensurable. There is no common or neutral language in which to express them. For example, consider the following pair of statements:

  Rabbit paradigm: "The distance from the mouth to the ear is X."
  Duck paradigm: "The distance from the mouth to the ear is Y."

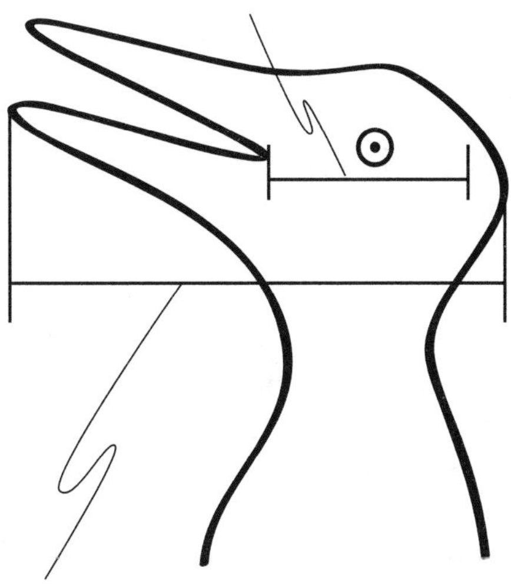

Figure 6 Kuhn's duck/rabbit.

They seem to contradict one another, but what each means by "ear" is different. The meaning of "ear" is paradigm dependent (see Figure 6).

Kuhn cites special relativity (1962/70, 101) as an instance of both incommensurability and of the noncumulativeness of the history of science. It's an example that can be appreciated by those with the appropriate technical background (others can skip it). Einstein and Newton agree that $F = d\mathbf{p}/dt$, but $\mathbf{p} = m\mathbf{v}$ in Newtonian mechanics while $\mathbf{p} = m_0\mathbf{v}/(1 - (v/c)^2)^{1/2}$ in special relativity.

Derivations to show that Newtonian mechanics is just a special case of relativity (i.e., at low velocity, when the ratio v/c goes to 0) often go like this:

$$\lim_{v/c \to 0} F_E = \lim_{v/c \to 0} \frac{d\, m_0 v / \sqrt{1-(v/c)^2}}{dt} = \frac{d\, m_0 v}{dt} = \frac{dmv}{dt} = F_N$$

Kuhn would claim that there has been a logical sleight of hand here. The illegitimate assumption is that relativistic rest mass, m_0, is the same thing as Newtonian mass, *m* (at least at low velocity), and that one can be substituted for the other in a derivation. These are quite different concepts, however, so the derivation is fallacious. Newtonian mechanics, Kuhn insists, is not a special case of Einsteinian mechanics.

- *Is paradigm change rational?* Many think that on Kuhn's account it cannot be. According to Imre Lakatos it's just "mob psychology" (1970, 178). Since methods, observations, assessments, and so on are paradigm dependent, it would seem there can be no neutral ground from which to judge the merits of rival paradigms. In a later essay, "Objectivity, Value Judgement, and Theory Choice" (1977b), Kuhn gives five criteria for change: empirical accuracy, consistency, broadness of scope, simplicity, and fruitfulness. These are trans-paradigm criteria. They make Kuhn's views seem more plausible, but are somewhat at odds with the more radical doctrines of *The Structure of Scientific Revolutions.*
- *Truth and progress:* Scientific realists believe that our best theories are (approximately) true–true in the sense of corresponding to an independent reality. Kuhn, however, flatly denies this. Not only our beliefs, but reality itself is paradigm dependent. With a paradigm, scientists "create" the world in which they work. Similarly, progress for Kuhn is not progress *to the truth* or to anything else; rather, it is progress *from*–there is no goal for science; it simply moves away from an earlier anomaly-ridden paradigm to a new one that is more empirically accurate, broader in scope, and so on. Kuhn cites an analogy: species in Darwinian evolution adapt to their environment and in this sense make progress, but there is no ideal form toward which a species is headed. Science is similarly driven from behind; it is not headed toward any goal such as truth.

Philosophers and historians of science have taken Kuhn to task on many fronts–though not before conceding the extent of Kuhn's extraordinary contribution. A brief review of some complaints is in order. For one thing, Kuhn's description of the historical record may be incorrect. He claims that normal science is dominated by a single paradigm, but the history of science seems to be full of co-existing rivals. For example, in physics: quarks vs. boot-strapping; in cosmology: big bang vs. steady state; in economics: Marx vs. Keynes; in psychology: Freud vs. Skinner, and so on. Nevertheless, the idea of a global unit that guides research is compelling. Another popular target is incommensurability. Kuhn (and others such as Feyerabend) have not made their case here. Their view rests on a particular account of language (involving contextual definitions, for example) that may simply be false. This remains an outstanding problem in the philosophy of science and philosophy of language. Much the same can be said about the theory-ladenness of observations. Again, the case has not been made fully and remains quite controversial. Nevertheless, the correct theory of observation will surely have to bow in Kuhn's direction.

Rationality has been the biggest sticking point for most of Kuhn's critics. As I said at the outset, if actual science is not a rational activity in the eyes of some normative philosophical account, then that should be taken as grounds for thinking the philosophical account is incorrect. Kuhn insists, especially in later writings (see *The Essential Tension*), that science *is* rational on his view. However, this remains a very contentious question.

Kuhn's Aftermath

Kuhn has had a tremendous effect. *The Structure of Scientific Revolutions* is one of the great books of the twentieth century. Even those who don't believe a word Kuhn has written should still concur with that claim. Theologians, social scientists, and literary critics, as well as historians and philosophers of science have all been struck by its central motifs. His book stimulated and promoted sociological approaches to science–much to Kuhn's chagrin. To a great many readers the switch from one paradigm to another did not seem to be grounded in "evidence" and "reason" as these are normally conceived; so any attempt to explain a shift in scientists' beliefs must appeal to some other type of cause, usually some sort of social factor. It was a small step from this to denying that reasons and evidence play any role whatsoever, or at least, any role that would be recognized by most working scientists, or philosophers of science, or traditional intellectual historians.

Social constructivists have found substantial ammunition in Kuhn's book. Let's remind ourselves of a few key passages.

> Like the choice between competing political institutions, that between competing paradigms proves to be a choice between incompatible modes of community life. Because it has that character, the choice is not and cannot be determined merely by the valuative procedures characteristic of normal science, for these depend in part upon a particular paradigm, and that paradigm is not at issue. When paradigms enter, as they must, into a debate about paradigm choice, their role is necessarily circular. (1962/70, 94)
>
> As in political revolutions, so in paradigm choice–there is no standard higher than the assent of the relevant community. To discover how scientific revolutions are effected, we shall therefore have to examine not only the impact of nature and logic, but also the techniques of persuasive argumentation within the quite special groups that constitute the community of scientists. (ibid., 94)
>
> In these matters neither proof nor error is at issue. The transfer of allegiance from paradigm to paradigm is a conversion experience that cannot be forced. (ibid., 151)

These are striking passages. And the way they strike most readers is as saying that nonrational factors play a decisive role in scientific decision-making. The earliest responses to Kuhn's book were from those who saw science as a rational activity and thought Kuhn's analysis fundamentally flawed. As mentioned earlier, Imre Lakatos called Kuhn's analysis a theory of "mob psychology" (1970, 178), and Dudley Shapere complained that paradigm change "cannot be based on good reasons" (1966, 67).

On the other hand, social constructivists would not be in the least upset with these alleged shortcomings. On the contrary, they might cheerfully agree with Bruno Latour, who remarked, "'Reason' is applied to the work of allocating agreement and disagreement between words. It is a matter of taste and feeling, know-how and connoisseurship, class and status. We insult, pout, clench our fists, enthuse, spit, sigh, and dream. Who reasons?" (1988, 179f). We can only wonder.

In discussing the positivists, Popper, and Kuhn, I have described three of the most important trends in twentieth-century philosophy of science. There are others. Besides the tremendous interest in social constructivism that was stimulated in part by Kuhn, there were (broadly speaking) two approaches to the philosophy of science that followed hot on Kuhn's heels.

Historical Approaches

The motivation for historical approaches stems from a basic Kuhnian precept: history can teach us a lot. The trick is to use history without abandoning rationality. Popper and the positivists are too *a priori* in their approach to the norms of science and do little or no justice to actual history. Rather than abandoning them completely, however, we still want an analysis that leaves science a basically rational activity. The leading figures in this approach include Imre Lakatos and Larry Laudan. There are others, but these two will suffice to illustrate the historical approach.

Lakatos's "methodology of scientific research programmes" (1970) posited a basic entity that is similar to Kuhn's paradigms, a *research program.* It is more general than any particular theory and is, in fact, a series of theories. A research program has a *hard core,* a set of fundamental principles that more or less define the program. It also has a *protective belt* that is constantly being modified and that takes the blame whenever something goes wrong. There are no crucial experiments, according to Lakatos, since we can always blame something in the protective belt, rather than blame the hard core. This point sharply separates him from Popper. Lakatos, an avowed Popperian, initially claimed to be proposing a version of Popper that had learned a thing or two from Kuhn. Popper would have none of it and claimed that Lakatos's views were fundamentally at odds with his own. There was, sadly, a professional and personal break between them before Lakatos's premature death in 1974.

Research programs that make the occasional correct novel prediction are *progressive,* according to Lakatos, whereas those that are constantly modifying themselves to keep pace are *degenerating. Rationality* in science consists in adopting progressive programs and abandoning degenerative ones. One should not jump ship too soon, but Lakatos, unfortunately, was never able to say how long a research program had to degenerate before it became irrational to stick with it.

Larry Laudan (1976) proposed the *research tradition,* and like Kuhn's paradigms and Lakatos's research programs, it is a global unit, not to be identified with a particular theory. A research tradition is *progressive* if it solves problems. There are two types: empirical and conceptual. An empirical problem might be: Why do the finches have such different characteristics from island to island in the Galapagos? A conceptual problem might be: Given the general correctness of a mechanical outlook, how can we explain the evolution of species without appeal to teleological causes?

Laudan distinguishes *accepting* from *pursuing* a research tradition. We accept something when we believe that it is true. We pursue a theory or tradition when we think it is promising and worth further investigation. *Rationality* in science consists in accepting the tradition that has solved the most problems and pursuing the one that is solving problems at the fastest rate. This simple distinction allows us to make sense of scientists, for example, in the early twentieth century who seemed to believe classical physics yet put their energy and resources into the new quantum theory. If one should only do research on what one believes to be true, then this behavior is puzzling, or even irrational.

Popper and the positivists would have been surprised if the great scientists had all turned out to be irrational, but it would not have bothered them, at least not in principle. Galileo, Newton, Darwin, and Einstein just might be irrational, after all. To champions of the historical approach, this is simply absurd. Lakatos and Laudan have produced their accounts of scientific method in light of the actual history of science; we might even say that they have tailored their accounts to match history. This raises a potential problem of circularity.

Lakatos and Laudan have given us an account of scientific method in which the history of science is rational, but how do we know they're accurate? They have assumed from the start that the history of science is rational. It would seem that we are begging the question against social constructivists, if we offer one of these historical approaches as evidence of the rationality of science.

The objection is fair, but it is not a crushing blow. Like many others, champions of the historical approach look at science prephilosophically and see a wonderful success story. They then see their job as trying to give a philosophically coherent account of how it works in detail. If one does not share this initial assessment of science, then, of course, one is not likely to be too impressed with the rationalistic accounts that stem from it. This does not mean that the initial judgment of the success of science is mistaken; it only means that the debate between the rationalists and the social constructivists will have to be settled on other grounds.

Logical Approaches

In probability theory, Bayes's theorem is not controversial. Yet many want to apply it in a setting of theory evaluation–and that is controversial. Let T be some theory and E be some piece of evidence. Then the claim is that Bayes's theorem gives us a way of estimating the probability of the theory, given the evidence. In the following formula, $P(T)$ means "the probability that T is true," $P(T|E)$ means "the probability that the theory T is true, given that the evidence E is true," and so on. $P(T|E)$ and each of the other terms is a number between 0 and 1.

$$\text{Bayes's theorem: } P(T|E) = P(E|T) \times P(T) / P(E)$$

Where do the probabilities on the right come from? These are known as the "prior probabilities." They must be filled in on the basis of background beliefs. If T implies E, then $P(E|T) = 1$, but generally there is only a statistical connection. For the most part $P(T)$ and $P(E)$ express subjective degrees of belief. For this reason, champions of the use of Bayes's theorem in theory evaluation are often called "subjective Bayesians." The actual degree of belief can be anything–it's purely subjective–but it must be coherent with the laws of logic and probability theory and with the other beliefs that are playing a role in the calculation.

The process of assessing the probability of a theory (sometimes called "Bayesian conditionalization") goes like this:

1. Assign the prior probabilities: $P(T)$, $P(E)$, $P(E|T)$
2. Use Bayes's theorem to calculate $P(T|E)$
3. Check to determine if E is true. (Suppose it is.)
4. Adjust your belief in T. It should now be $P_{new}(T) = P(T|E)$.
   In other words, drop $P_{old}(T)$ and adopt $P_{new}(T)$

In the last chapter I cited examples of novel predictions. Those are the sorts of examples that fit the Bayesian pattern. We could look briefly at the Fresnel case to illustrate. Suppose we assign numbers as follows, then calculate what our confidence in the theory should be in the light of the experimental outcome.

$P(E|T) = 1$ (since Fresnel's wave theory implies the bright spot)
$P(T) = \text{Prob}(\text{Fresnel's theory}) = 0.1$ (its subjective prior probability is very low)
$P(E) = \text{Prob}(\text{there will be a bright spot}) = 0.11$ (its prior probability is also very low, but consistency requires that it not be lower than $P(T)$, since T implies E)
$P(T|E) = P(E|T) \times P(T) / P(E) = 1 \times 0.1 / 0.11 = .91$
Determine experimentally that E is true.
Revised belief: $P_{new}(T) = .91$

Given these numbers (which are rather arbitrary), our rational belief in Fresnel's wave theory goes from a low 0.1 to a high .91 in the light of the experiment that revealed the unexpected bright spot.

The Bayesian approach is very impressive and has many champions. It also has problems. How can we justify using prior probabilities when they are so subjective? What sense are we to make of so-called "old evidence"? Unlike the bright spot in the Fresnel case, evidence is often known in advance of any theory that explains it. Logical omniscience is required and this seems a hopeless idealization. Scientists, of course, can't see all the logical consequences of their beliefs, and contradictory assignments of prior probabilities will make the application of the theorem unworkable. Finally, what has Bayes's theorem to do with actual science, since working scientists don't reason like this anyway?[3]

I mention these post-Kuhnian developments to indicate that philosophy of science didn't stop with Kuhn. It's flourishing in several directions, and there are many options and suboptions even within the historical and the logical approaches. The crucial commonality is that they each see science as a rational activity. Members of both the logical and the historical camps oppose social constructivism. Instead of trying to spell out in greater detail current views in the philosophy of science, it would be much better to focus on a few central notions that are bandied about. Three of the most important are: *realism, objectivity,* and *values.* A later chapter will be devoted to them.

4

The Nihilist Wing of Social Constructivism

The more memorable attacks by Sokal and others tend to be attacks on so-called postmoderns, the more nihilist wing of social constructivism. They are a special breed within the broader social constructivist camp. What makes them particularly susceptible to lampooning? Postmodernism stands in opposition to the Enlightenment, which is taken to be the core of modernism. Of course, there is no simple characterization of the Enlightenment any more than there is of postmodernism, but a rough and ready portrayal might go like this: Enlightenment is a general attitude fostered in the seventeenth and eighteenth centuries on the heels of the Scientific Revolution; it aims to replace superstition and authority by critical reason. Divine revelation and holy scripture give way to secular science; tradition gives way to progress. Enlightenment advance is of two sorts: scientific and moral. Our scientific beliefs are objectively better than before and are continuing to improve, and our moral and social behavior is also improving and will continue to do so.

There is another aspect to modernism that is often linked to the Enlightenment, but seems to go well beyond it. This is the doctrine that there is one true story of how things are. Jean-François Lyotard, one of the most prominent postmodern commentators, speaks of the "incredulity about metanarratives" (1984). Science for him is just a game with arbitrary rules, and truth is nothing more than what a group of speakers say it is. Although the Enlightenment figures that postmoderns attack would have happily embraced the view that there is one true story (perhaps with qualifications), so would Aristotle and so would the medievals. In attacking so-called "grand narratives" or "metanarratives," postmoderns are attacking much more than modernism.

Roman Catholicism's fondness for tradition and authority may stand in opposition to the Enlightenment, but that religion certainly disdains relativism and embraces the idea of a single truth.

Just as critical reason is seen by postmoderns as a delusion, so are all attempts to generalize or universalize. In place of so-called "totalizing" accounts of nature, society, and history, "local" accounts are offered. *Localism* or *perspectivalism* is the view that only very limited accounts of nature, society, or whatever the subject of discourse are to be taken as legitimate; grand theories are invariably wrong, or oppressive, or both.

Jacques Derrida, another leading postmodern figure, has pronounced that any attempt to say what postmodernism is (or what it is not) will invariably miss the point. This puts the would-be expositor at a serious disadvantage. Nevertheless, it seems reasonably fair to say that three ideas are central to postmodernism: one is the *anti-rationality* stance; a second is the *rejection of objective truth;* and the third is *localism.*

The naturalistic versions of social constructivism (to be discussed in a later chapter) overlap partially with postmodern accounts, but they tend to stem from quite different motivations. As we shall see, Bloor and his Sociology of Scientific Knowledge (SSK) colleagues start from science and arrive at an analysis in which the social plays a dominating role. On the other hand, postmodern analyses stem from recent political and historical considerations. World War I, for instance, shook the faith in rationality and progress of those who matured in the nineteenth century—a faith that Marx shared in full with the Victorian bourgeoisie. Those who hadn't abandoned their Enlightenment hopes as a result of this senseless catastrophe had only to wait a few more years for Hitler and the Holocaust. In his famous "What is Postmodernism?" Lyotard writes:

> The nineteenth and twentieth centuries have given us as much terror as we can take. We have paid a high price for the nostalgia of the whole and the one . . . we can hear the mutterings of the desire for a return of terror, for the realization of the fantasy to seize reality. The answer is: Let us wage war on totality . . . (1983, 46)

Localism, it would seem, is proposed as an antidote to war. As for the Holocaust, modernity itself is responsible, according to Zygmunt Bauman, who speaks of "the evil of rationality." (1989, 202f)

Although postmodernism is a recent intellectual phenomenon, its roots are deep. Friedrich Nietzsche, more than a century ago, announced that all thinking is perspectival, that there are no facts–only interpretations. Nietzsche's nihilist attitude toward truth and knowledge has been a primary source of inspiration to postmodernism. Yet one might wonder why Nietzsche has been so popular. The answer seems to be that in spite of his rejection of all truths and all values, in spite of his belief that the cosmos is profoundly indifferent to our aspirations and opinions, Nietzsche still finds the strength to say "yes" to life. The struggle of the *will* in the face of meaningless chaos is what readers find intriguing. This aspect of Nietzsche lives on in postmodernism. We do not discover how things are, we instead impose our wills on formless nature.

No one would call Nietzsche part of the "academic left." But the attraction of such a view should be apparent. If "reality" isn't a thing in itself waiting to be discovered, but rather is something to be constructed by our wills, then let's structure it in a democratic way that suits all of humanity equally. Nietzsche, himself, wanted it left to the natural aristocrat, the "superman." This is the right-wing side of social constructivism, a doctrine admired by Hitler. Given the Nietzschean background, the science wars (in the view of postmoderns) then become a fight over how we are to make or re-make the world we live in.

A Too-Easy Refutation

In a widely read piece praising Sokal, Paul Boghossian (1996) uses a self-refutation argument against postmodernism. Boghossian is a paradigm analytic philosopher, well known for his work in the philosophy of language and related topics. He exemplifies many of the genuine strengths (and weaknesses) of analytic philosophy. Here, for instance, is his quick rebuttal of postmodernism:

> If a claim and its opposite can be equally true provided that there is some perspective relative to which each is true, then since there is a perspective–realism–relative to which it's true that a claim and its opposite cannot both be true, Postmodernism would have to admit that it itself is just as true as its opposite, realism.

> But Postmodernism cannot afford to admit that; presumably, its whole point is that realism is false. Thus, we see that the very statement of Postmodernism, construed as a view about truth, undermines itself; facts about truth independent of particular perspectives are presupposed by the view itself. (Boghossian 1996, 15)

There is a sense in which Boghossian is perfectly right–postmodernism undermines itself. But this kind of quick rebuttal fails to do full justice to the postmodern vision. Every beginning philosophy course introduces the problem of skepticism. And the bright sparks in the class immediately offer the following refutation: "Skeptics say that no belief is justified; thus, their claim is itself not justified; consequently, there is no reason to be a skeptic. QED."

In some strict but uninteresting sense the postmodern and the skeptic are refuted by these arguments. But such refutations are insensitive to the problems at hand and the nature of the claims being made. It may well be that the postmodern and the skeptic are right, but simply lack the linguistic or conceptual resources to state their claim in any but a paradoxical way. A much fairer reading might be in the spirit of the early Wittgenstein. I mentioned this example in a previous chapter, but it's worth repeating here. In his *Tractatus,* Wittgenstein offered a theory of meaning and truth. The consequence, which he fully recognized, is that all the statements in the *Tractatus* itself are meaningless. His own theory was one of those things that, by his own lights, could be shown, but not said. In other words, the *Tractatus* itself is, strictly speaking, nonsense. Wittgenstein introduced his now famous metaphor: reading and understanding the *Tractatus* is like climbing a ladder; after we have climbed to the top, we can throw the ladder away, for now we see the world aright.

This, I think, is a much fairer way to view the situation. If we are to refute skepticism or postmodernism, it will have to be some other way, a way that pays as much or more attention to the practices of postmodernism and the spirit behind its doctrines as it pays to the specific sentences that are uttered. When Derrida claimed that any description of postmodernism misses the point, he was not just weaseling his way about, but actually had some warrant–though he was certainly weaseling, too.

An Irritating Style

A more charitable approach to postmodernism would focus less on the explicit form of pronouncements, which may be unfortunate, and more on the content within and the practice behind them. To this end, let's look at a particular article by Stanley Aronowitz, "The Politics of the Science Wars" (1996). This article appeared in the same infamous issue of *Social Text* that contained Sokal's hoax, so it's a natural subject for consideration. Aronowitz is a prominent and influential player in cultural studies of science. The case he makes in *Social Text* includes bringing "science down to earth, to show that it is no more, but certainly no less, than any other discourse. It is one story among many stories that has given the world considerable benefits including pleasure, but also considerable pain" (1996, 192).

Aronowitz begins by locating matters (and his allegiances) in the larger social situation. "The main fronts of the Culture Wars–Western civilization versus multiculturalism, high versus low in music, literature, art, and modernism versus Postmodernism–are now joined by what may be termed the Science Wars" (1996, 177). He goes on to jibe, "Members of the faith are circling the wagons against what they perceive to be a serious threat to the church of reason," then adds, "While everybody, including politicians and molecular biologists, is qualified to comment on politics and culture, nobody except qualified experts should comment on the natural sciences" (ibid., 178).

Remarks such as these might be overblown, but they are still fair comment and part of the normal rhetoric of debate. More irritating is Aronowitz's habit of making major claims that are not explained or justified beyond citing some other author, invariably another author whose views are highly controversial. Thus, he says that science is not "a steady march toward the Truth (Mulkey; Barnes)," then he says that science is "conditioned by these [social] circumstances (Bloor)," "scientific knowledge is itself 'gendered,'" and "The key players and their institutions are the recognized gatekeepers of what counts as science and, more broadly, what counts as truth (Birrer)" (Aronowitz 1996, 179ff). On and on he goes in this vein.

This rhetorical practice is on a par with my writing, "As Newton has shown, science is wonderful, and as Darwin proved, science is great, and as Einstein amply demonstrated, science is terrific." The Aronowitz argument style is little more than name-dropping. If there is a case to be made, he must make it with argument, not with endless appeals to authority. Background information can be justified with a footnote, but central, contentious claims need more justification than a remark of the form: "As the prophet has said . . ."

Aronowitz often talks about some of the details of science in a way that can only irritate the knowledgeable while leaving nonscientist readers baffled, or worse, intimidated. Consider the following statement, which he says most physicists (unlike himself) believe. "One day scientists shall find the necessary correlation between wave and particle; the unified field theory of matter and energy will transcend Heisenberg's uncertainty principle" (ibid., 181). I doubt that any working physicist believes anything of the sort. Insofar as it is intelligible at all, it's simply false. (But I can't be sure since I have no clear idea what "necessary correlation between wave and particle" could mean.) As for the Heisenberg principle, that's quite central to any quantum theory, including any future unified field theory. Like many commentators on quantum mechanics, Aronowitz does not seem to understand that filling the gap in the uncertainty principle is, in effect, rejecting the theory, not just finding another piece to the puzzle. No one who accepts quantum mechanics believes that the uncertainty principle can be transcended.[1]

This muddled passage is preceded by another that is both wrong and highly misleading. After applauding the "critical theory of science," Aronowitz contrasts it with "orthodox positivism," for which the "historicity of science must be confined to the idea that scientific theories are, at best, successive approximations of a *reality existing outside the conditions of investigation*" (ibid., 180f, my italics). The passage is incorrect because positivism has traditionally *denied* that there is any reality independent of knowers. Positivism–in contrast to social constructivism–is arguably an objectivist view of human knowledge, but it is certainly not a form of realism (see the discussion in the next chapter).

Even worse, the passage is highly misleading, because it suggests that positivism is the *alternative* to the view that Aronowitz urges. Kuhn's famous account of science in *The Structure of Scientific Revolutions* (to cite another view) is completely "internal"; that is, the history of science is driven by scientific factors, not social ones. Yet Kuhn absolutely denies that there is any truth that successive theories or paradigms are approximating. Larry Laudan, to cite yet another example, staunchly opposes almost everything Aronowitz endorses, but he, like Kuhn, is adamant that there is no ready-made world that we can with confidence truly describe (Laudan 1984). Much of the appeal of social approaches to science is based on caricatures of its rivals.

Aronowitz misinterprets Kuhn's views on truth: he says, "Kuhn remains agnostic with respect to the truth value of the new paradigm but, tacitly, accepts a highly relativistic version proposed by Peirce: the truth is what those who are qualified say it is" (1996, 191). This is a travesty of both positions. Peirce, like any sophisticated verificationist, distinguished *truth* from *what is currently believed to be the truth,* even if it is believed by all the experts. Truth for Peirce is what all scientists agree upon at the "end of inquiry," an idealized notion, never to be reached in any finite time.[2] This distinction between truth and what is currently believed to be true is key to the objectivity claimed by verificationists who reject any realist conception of truth. It is also what sharply separates them from those like Aronowitz, who see truth as nothing more than a transient social consensus.

One of Aronowitz's major claims in his paper is that the natural sciences and the social sciences should be treated alike. In this he is opposing a common feature of much social constructivism. Often accounts of the natural sciences are treated as constructions, but explanations of these constructions in terms of various social factors are invariably treated as objectively true. (This theme will arise again in later chapters.) Thus, what the Weimar scientists believed about the atom or what Pasteur believed about spontaneous generation are taken to be constructions that served their respective social interests. But the claim that they actually had those interests (respectively, in regaining lost prestige or in defending the Second Empire) is taken to be literally, objectively true. (I'll describe these now famous examples in Chapter 6.)

Such a position has seemed highly unsatisfactory to opponents of social constructivism and even to some of its practitioners, such as Latour. Aronowitz is not addressing this specific point, but he is concerned with the more general issue of the asymmetry of the natural and the social. After reviewing and rejecting some of the traditional reasons given for separating the social from the natural sciences, Aronowitz offers the following as an argument for treating them alike.

> My argument . . . depends on one of the more important interpretations of quantum theory . . . Bohm's solution to the posit of a split between knowledge and its object, which dominated classical physics and remains among the most influential interpretations of quantum mechanics, is to argue for "the *undivided wholeness* of the measuring instrument and the observed object." . . . For Bohm and others who attempt to overcome the dualism of the observer and the observed, the field is not constituted by objects whose antinomy is the subject-observer (ibid., 185).

There are several points to take up here. But first let's note that Aronowitz's attempt to join the social and the natural sciences is the reverse of traditional attempts. Usually, the social sciences are assimilated to the natural. Aronowitz wants things to go in the opposite direction: the natural sciences (at least quantum mechanics *à la* Bohm) are supposed to be just like the social sciences. I won't attempt to evaluate Aronowitz's claim; rather, I'll turn to the details of his argument.

Aronowitz rather badly misunderstands Bohm's view. The key, as he sees it, is how Bohm is able to "overcome the dualism of the observer and the observed." Bohm's holism does no such thing, at least not in the sense that Aronowitz has in mind. Let me try to explain with an analogy. A physical property such as *mass* is *intrinsic* to an object; but *weight* is a *relational* property. A block of butter will have the same mass here as on the Moon, but its weight changes considerably. Weight depends on the relation it has with Earth or with the Moon. The gravitational force is different in the two cases. However–and this is the crucial point–the relational property of weight is just as objective and observer-independent as the intrinsic property of mass. Of course, social constructivists would state that neither is objective and observer-independent, but relationalism (or holism, if you like) has nothing to do with it. Now to Bohm.

Bohm offered a so-called contextual hidden variable theory of the quantum world. The contextual part of Bohm's theory is this: the electron possesses properties only in relation to the rest of the universe, including the measuring apparatus. Properties such as position, velocity, and spin were treated as intrinsic but objective properties in classical physics, then as intrinsic but nonobjective properties in standard interpretations of quantum mechanics (e.g., the Copenhagen interpretation). Bohm has made them objective but nonintrinsic. This is Bohm's idea of holism, contextualism, or relationalism. It has nothing special to say about observers; human consciousness plays no role. Bohm's theory certainly will not serve Aronowitz's purposes in any conceivable sense.

The misunderstandings of science are all too common, but it would be a serious mistake to think it takes ideological commitment to corrupt an otherwise clear-headed grasp of the facts. Here's a comical example of confusion, taken from a recent Internet conversation.[3]

A: A pen always falls when you drop it on Earth, but it would just float away if you let go of it on the Moon.
B: What? A pen would *fall* if you dropped it on the Moon, just more slowly.
A: No it wouldn't, because you're too far away from Earth's gravity.
B: You saw the APOLLO astronauts walking around on the Moon, didn't you? Why didn't they float away?
A: Because they were wearing heavy boots.

Postmodern Language

Postmodernists have been heavily criticized for their language. The common complaint is that it is hopelessly obscure. This is not a charge that could be laid at the feet of Aronowitz, who is often if not always admirably clear. But it is heard often enough and one wonders how fair it might be.

In general we allow that any discipline has its technical terms that are highly useful for expressing the special concepts of that field. Some training will be required to understand a theoretical physicist who, for example, states the first of Maxwell's laws of electrodynamics: "The flux of an electric field through any closed surface is equal to the net charge inside divided by the permittivity of free space." Expressed that way, it is hard enough to understand; when the usual symbols are used, $\nabla \cdot \mathbf{E} = \rho/\epsilon_0$, all pretense at comprehension by the uninitiated must be abandoned. We have to learn the language, and it's going to be a lot of work.

Is postmodern literature any different? Many of the repeatedly used terms must be mastered. "Difference," "discourse," "essentialism," and "fractured identities" are technical terms that have to be learned just as "divergence of a vector field" has to be grasped in order to comprehend classical electrodynamics. Postmoderns can't be blamed for this problem with their language. It's unfair that they are.

Yet it isn't only the jargon itself that frustrates the average reader. In reading postmodern literature, one gets the feeling that the language is used in a very self-conscious way. Scientists are self-conscious, too, in their use of language; technical terms are carefully defined and carefully employed. There is an important difference, however. Scientists use technical terms for the sake of precision and economy; postmoderns draw attention to the words themselves. This could be a valuable activity. Postmodern authors often deliberately use normal words in an abnormal way. The point is to jar readers out of any complacency about the concepts they are taking for granted. Thus, instead of "gynecology" one reads "gyn-ecology"; instead of "malpractice" we have "male-practice"; "dis-ease" for "disease," "alien-nation" in place of "alienation," and so on.

Language, according to the postmoderns, isn't representational; words don't hook onto objects in any straightforward way. Postmodern prose is sometimes deliberately unpleasant for good reason–it's didactic. It's designed to jar us out of thinking instinctively in fixed representational ways. If we take the postmoderns' point about language to be valid, then they can't be faulted for their writing style. The obscurity is (at least sometimes) deliberate and it could serve a good end.

There is also a kind of playfulness in postmodern discussions, a playfulness that one might expect from literary critics and others whose professional lives are devoted to language. Writing in the same issue of *Social Text* that contained the Sokal hoax, and commenting on a passage from a book by Andrew Ross, George Levine says, "books written like this depend on readers who like to play with language, but they are often . . . read by people who don't" (Levine 1996, 123). He goes on to suggest that irony and metaphor are lost on scientist readers. We can overlook this slight for now, but consider his underlying point: There are writers who aim at the truth and writers who aim at something different. Of course, irony and truth needn't be mutually exclusive. Being truth-oriented, however, implies that one ultimately wants to translate ironic or metaphorical language into something literal; irony and metaphor are seen as useful, witty tools, not ends in themselves.

Is this the whole story? Even after taking legitimate postmodern concerns with language into account, still too often the ratio of empty verbiage to real content is vanishingly small. Noam Chomsky (responding to postmodern attacks on science and reason) offered this rough test:

> I have spent a lot of my life working on questions such as these [the origins of human knowledge or the nature of U.S. foreign policy] using the only methods I know of–those condemned here as "science," "rationality," "logic," and so on. I therefore read the papers with some hope that they would help me "transcend" these limitations, or perhaps suggest an entirely different course. I'm afraid I was disappointed. Admittedly, that may be my own limitation. Quite regularly, "my eyes glaze over" when I read polysyllabic discourse on the themes of poststructuralism and Postmodernism; what I understand is largely truism or error, but that is only a fraction of the total word count. True, there are lots of other things I don't understand: the articles in the current issues of math and physics journals, for example. But there is a difference. In the latter case, I know how to get to understand them, and have done so, in cases of particular interest to me; and I also know that people in these fields can explain the contents to me at my level, so that I can gain what (partial) understanding I may want. In contrast, no one seems to be able to explain to me why the latest post-this-and-that is (for the most part) other than truism, error, or gibberish, and I do not know how to proceed. (Chomsky, *Z-Net*)

Chomsky then reiterates a point often made by others. There are particular uses of science and reason that have actually been abuses; reason has been perverted for some harmful end. It is these perverted uses of science that are quite legitimately attacked by postmoderns and others, not reason itself.

> The critique of "science" and "rationality" has many merits, which I haven't discussed. But as far as I can see, where valid and

> useful the critique is largely devoted to the perversion of the values of rational inquiry as they are "wrongly used" in a particular institutional setting. What is presented here as a deeper critique of their nature seems to me based on beliefs about the enterprise and its guiding values that have little basis. No coherent alternative is suggested, as far as I can discern; the reason, perhaps, is that there is none. What is suggested is a path that leads directly to disaster for people who need help–which means everyone, before too long. (ibid.)

Feyerabend's Development

Most of those in the nihilist camp have little training in and even less understanding of the sciences. But there are important exceptions. One of these is Paul Feyerabend, an iconoclast of the first rank. Feyerabend's early education was in physics in Vienna. After serving in the *Wehrmacht* during the war (in which he was seriously wounded), he became for a while a follower of Popper. His academic career unfolded in many places, including London, Berkeley, and Zurich. Feyerabend's early work in the philosophy of science (in the late 1950s and early 1960s) was brilliant, but still within the orthodox camp. His views grew increasingly radical over the years so that by the 1970s and after he was writing books with titles such as *Against Method* (1975) and *Farewell to Reason* (1987). Even though we may ultimately reject it as incredible, Feyerabend's nihilism developed in a natural and plausible way.

Feyerabend and Kuhn were close colleagues at Berkeley in the early 1960s, and on the basis of numerous conversations developed a common skeptical opinion of the received view of science. Kuhn, of course, wrote his famous book at this time. Less well known, but just as brilliant is Feyerabend's essay from this era, "Explanation, Reduction, and Empiricism" (1962). Among other things, it showed how tricky the notion of evidence is.

Suppose you have a theory of how cancer cells come into existence, how they develop, how they can be destroyed, and so on. What if I come along and say: "Well, that's all very good, but you haven't explained why the sky is blue or why the stock market fell yesterday." You would very likely be irritated by my silly remarks, since you would think my requests irrelevant. Your theory is about cancer, not about the sky's color or the stock market's direction; so you don't have to explain those things and your theory is in no way diminished by not being able to do so.

This seems a perfectly reasonable response. Let's make it a bit more precise, filling in a few gaps. We could designate a phenomenon by a pair of statements, $\{P, \sim P\}$. For example, the phenomenon of the color of the sky is represented by the pair: {The sky is blue, The sky is not blue}. When a theory (perhaps in conjunction with other background information) implies one statement or the other (i.e., $T \rightarrow P$ or $T \rightarrow \sim P$), then we can say that the phenomenon in question is relevant to the theory T. Thus, "Cancer cells flourish in condition C" is evidentially relevant to your theory T provided that either T implies this statement or T implies its negation. Since your theory implies neither "The sky is blue" nor its negation, we can say that the color of the sky is evidentially irrelevant to T.

This is just a rough and ready characterization of the problem, but it will do for now. It's close to common sense and something like it is probably held by most working scientists. With a brilliant example, Feyerabend showed how hopelessly wrong this sort of conception of *relevant evidence* really is. His example concerns the rivalry between classical thermodynamics and the kinetic theory of matter and heat; it's worth working through.

Classical thermodynamics involves such concepts as pressure, volume, and temperature. These are all more or less observable notions, so it is not surprising that the theory was often called phenomenological thermodynamics. Its two central principles are: the first law of thermodynamics, which says that heat is a form of energy and energy is conserved; and the second law, which says that in any change of state the entropy will either remain the same or increase. (We might take it as also denying that heat and pressure are to be understood as involving the motion of tiny material particles.) The theory was a great success and its explanatory scope and power were very impressive.

The rival kinetic theory posited a hidden realm of tiny bits of matter that move about obeying Newton's laws. Heat, according to this theory, is nothing but the energy of motion at a micro-level, and pressure (on the wall of a container) is nothing but the bombardment of the tiny particles against the wall. Since Newton's laws also demand the conservation of energy, the first law of thermodynamics was easily covered. But the second presented no end of trouble. The best the kinetic theory could offer was an approximation based on probabilistic considerations. Champions of thermodynamics might well wonder why they should switch from their completely adequate theory when the new theory could only approximate the second law, a law that seemed completely inviolable.

Early in the nineteenth century a Scottish biologist, Thomas Brown, noticed something remarkable. The tiny bits of pollen he was examining under a microscope continued to move rapidly about in spite of his attempts to stop them. Were they alive? He'd wait for them to die. Were they feeding off something in the water in which they were suspended? He'd use distilled water. Was the table on which he was working vibrating? He'd make it more sturdy. Nothing worked; he couldn't eliminate the perpetual, random movement, now called Brownian motion.

Should proponents of classical thermodynamics have worried about Brownian motion? Champions of that theory would no more think about Brownian motion than they would think about the stock market or the rabbit population in Australia. Of course, Brownian motion is interesting and champions of classical thermodynamics would have been glad to know what was going on, but there was no particular reason to think that it had anything to do with them. Brownian motion is not in the domain of classical thermodynamics.

All that changed early in the twentieth century. Einstein–who wasn't really aware of Brownian motion in 1905–pointed out that the kinetic theory implied that tiny but still observable particles would be knocked around by fast-moving molecules, and that the paths of these tiny particles could be perceived. Others immediately realized that the phenomenon Einstein predicted is none other than Brownian motion. Later work by Perrin and others showed that the predictions of the kinetic theory were very accurate. Brownian motion thus became a spectacular triumph for the kinetic theory of heat and matter.

Promoters of classical thermodynamics had cheerfully ignored Brownian motion just as they had ignored eclipses, comets, and the phases of the moon, but they could ignore it no more. Suddenly classical thermodynamics was obliged to account for Brownian motion, which passed from being an evidentially irrelevant phenomenon to being very relevant indeed.

The philosophical moral is simple: *Theory testing is comparative.* We do not take a theory directly to nature to test its correctness. Instead we test a theory comparatively in the sense that we look to both nature and its rivals. A phenomenon becomes evidentially relevant to some theory when a rival theory can account for it.

The contrast with Popper and others couldn't be greater. According to Popper, a theory will tell you what relevant evidence for testing that theory is: it's the sum total of all the observable consequences of that theory. Feyerabend's wonderful example shows that there are more than just logical consequences that can play a role in supporting or refuting a theory.

There is another moral that is derived from the first, though Feyerabend did not stress it until years later–*pluralism.* We ought to cultivate more rival theories, since rivals can spawn more relevant evidence. Without the kinetic theory we would have never seen the shortcomings of classical thermodynamics. No amount of direct testing would have sufficed to make its problems manifest. There's a picture of science that goes like this: When you find out that you are wrong, go back to the drawing board. The rule might be better turned on its head: No matter how good your theory seems, go back to the drawing board anyway and construct more alternatives; that may be the only way to discover that your initial theory is wrong.

Feyerabend's analysis of this example (including the morals he drew from it) is one of the masterpieces of philosophy of science. It was an important step on his way to greater radicalism. Before examining the rest of Feyerabend's career, two things should be noted about his account of the thermodynamics example. First, it shows the utility of a liberal view of diversity. It's not just polite or democratic to let others do their own thing–it sometimes pays off in better science. Second, accumulating a diverse group of theories is still completely within the rationalist camp–we choose the best theory on the basis of evidence, not on the basis of social factors.

By the 1970s Feyerabend's radicalism had grown considerably. For the most part it stemmed from the pluralism implicit in the analysis just cited. He came to describe himself as an "epistemological anarchist" and a "dadaist,"[4] and as far as scientific method is concerned, he declared: "Anything goes." Feyerabend came to believe that scientific method (any prescribed set of rules) is at best sterile and often downright harmful in the development of science. He repeatedly cited historical examples to show that great science often broke the rules, and he claimed that people like Galileo were rhetorically brilliant–liars and cheats. Far from condemning such activity, Feyerabend described it as part and parcel of good science. Strict methods would get in the way of creative science, hence his anarchism, reflected in the title of his most famous book, *Against Method* (1975).

Feyerabend was quick and keen to draw political morals from this pluralism. "A free society is a society in which all traditions have equal rights and equal access to the centres of power" (1978, 9). In his later writings this became a dominant theme. Feyerabend believed that competing theories, whether they were drawn from regular scientific traditions (e.g., the tradition of classical mechanics) or what we call pseudo-scientific traditions (e.g., astrology), enriched people's lives and promoted a healthy habit of questioning our own views. It is not some pretentious notion of "truth" that really matters, says Feyerabend, it is human well-being.

Feyerabend's case for anarchy is based on a number of considerations–all rather weak. He takes common examples of methodological pronouncements, then shows that there are numerous examples in the history of science where those rules were clearly violated. Rules such as "Don't accept logically inconsistent theories" or "Don't accept theories that are in conflict with observation" have been breached by all sorts of great scientists. Looking at the course of history, these methodological infractions are happy events. If Galileo or Darwin or Einstein had stuck to the orthodox rules, we would be the worse for it.

One of the problems with Feyerabend's case is that the rules he chooses to subvert are very likely not the sort of rules that any serious methodologist would propose, anyway. Most philosophers of science who write on these matters typically make a distinction between rules that guide our *beliefs,* what we actually accept as true, and rules that guide our creative work in constructing or modifying promising lines of inquiry. I don't for a moment believe in precognition or other forms of ESP–theories that are for the most part incoherent and empirically inadequate. Nevertheless, I could easily be persuaded to sink a small amount of research money into its further study. If it turned out to be true, the payoff for this small wager would be great.

We can all agree that a true theory cannot be logically inconsistent; if it were, it wouldn't be true. Such a theory shouldn't be believed. But there may be a number of features of such a theory that seem promising. If so, then the theory–incoherent though it is–is worth the time and effort that might be needed to make it better. (The "old quantum theory" of Bohr and Sommerfeld is an example.) Exactly the same statement can be made about getting the observable facts right. A theory that failed shouldn't be believed, but it might still have enough promise that more time and money should be spent on its development.

Quantum electrodynamics is a perfect example. If it is not now, it certainly was plagued at one time with serious inconsistencies. When calculating the so-called self-energy of the electron, the answer was infinite. Still, the theory seemed to be on the right track in so many other ways. Eventually workable modifications (a process called renormalization) eliminated the logical problems, so that now QED, as it is commonly known, is a great success story. No one should have believed the theory in its early days; but thinking it promising and worth the effort was perfectly sensible. No one ever thought a contradiction was true (i.e., no one thought the self-energy of the electron is both finite and infinite); so no one could be said to have abandoned the rule "Don't accept contradictory theories." Feyerabend has a misleading tendency to tell stories like this one in which people seem to accept contradictions. A simple distinction between *theories with evidence for their truth* and *theories that seem promising* makes all the difference. Most new theories fall into this latter class and need to be protected temporarily from the kinds of standards we demand any acceptable theory eventually pass. Much of Feyerabend's persuasiveness stems from blurring what is implicitly evident to most working scientists.

Though Feyerabend's later anarchistic pronouncements are far-fetched, his earlier conclusions are often sound. The moral for us is the one he himself drew about theory pluralism–the more the merrier.

The "Postmodern Sciences"

There is an interesting distinction to be made between *postmodern accounts of science* (which are invariably anti-objectivist) and *postmodern science.* With surprising frequency postmoderns cheerfully embrace a number of particular sciences. Quantum mechanics (or at least aspects of it) and chaos are favorites. It is easy to see why. Heisenberg's uncertainty principle (according to a common interpretation) suggests that we humans are not merely observing the world in a passive way, but are somehow actively involved in making it what it is. Chaos theory makes the world out to be a wildly unpredictable and uncontrollable place. These feed into popular postmodern themes about the subjectivity, contingency, instability, and complexity of the world.

When critics of the postmoderns score direct hits, it is usually on this topic. It is somewhat ironic that the postmoderns are most vulnerable when they are actually approving of something they take to be part of current science. The problem invariably is a terrible misunderstanding of the particular science in question. Gross and Levitt, who seldom err on the side of generosity, are probably right in saying these accounts are "shallow and confused, and apparently arise from a botched reading of popularizations" (1994, 97). They certainly cite lots of examples. One postmodern advocate, Steven Best, distinguishes the "linear" equations of Newton and quantum mechanics from the "nonlinear" equations of chaos theory. As Gross and Levitt rightly point out (ibid., 98), Newtonian celestial mechanics is highly nonlinear and is where chaos got its start (in the work of Poincaré).

Most of Sokal's best jokes are targeted at postmodernists' botched interpretations of accepted science, such as Aronowitz on quantum mechanics or Hayles on hydrodynamics. In the opening chapter I quoted two excerpts and asked readers to identify which was nonsense. The nonsense piece constructed by Sokal was aimed at remarks by Jacques Lacan that try to use topology to interpret the mind. In a passage of mind-numbing bizarreness Lacan had identified the psychological subject with a Möbius strip.

Katherine Hayles's grip on recent intellectual history is similarly tenuous. She describes logical positivism as flourishing in the nineteenth century, although its heyday was most definitely in the first half of the twentieth, between the two wars. She describes special relativity and quantum mechanics as having undergone "substantial modification" as a result of their union in the creation of quantum field theory, which is now "played out." This is completely incorrect, as Gross and Levitt point out (ibid., 101). Special relativity and quantum mechanics have remained very stable since their inceptions in 1905 and 1925, respectively. And quantum field theory is currently a thriving research program.

There is an easy condemnation of postmodernism that we should resist. If postmoderns are happy to accept some results of current science, aren't they falling into a contradiction? On the one hand they deny the objectivity of science, but on the other they seem to accept the objectivity of quantum mechanics or chaos. Isn't this absurd? No. Their position is consistent, even if no one actually holds it: Science is not objective, but some nonobjective theories such as quantum mechanics and chaos are more congenial to progressive social ends. So, they could argue, we should adopt those theories. This may not be a plausible view, but it is consistent. Still, the implausibility remains a serious problem. First, it has not been shown that science fails to be objective, and second, the particular interpretations of quantum mechanics and chaos embraced by postmoderns are hopelessly confused.

Who is responsible for this silliness? Doubtless there are many sources, but physicists themselves must shoulder part of the blame, and the historian of physics Mara Beller rightly wonders "At whom are we laughing?" (1998). Although she enjoyed the Sokal hoax, she notes that much postmodern misunderstanding of modern science is taken from the popular pronouncements of some of the greatest physicists. First, from Max Born:

> The thesis "light consists of particles" and the antithesis "light consists of waves" fought with one another until they were united in the synthesis of quantum mechanics . . . Only why not apply it to the thesis Liberalism (or Capitalism), the antithesis Communism, and expect a synthesis, instead of a complete permanent victory for the antithesis? There seems to be some inconsistency. But the idea of complementarity goes deeper. In fact,

> this thesis and antithesis represent two psychological motives and economic forces, both justified in themselves, but, in their extremes, mutually exclusive. . . . There must exist a relation between the latitudes of freedom Δf and of regulation Δr, of the type $\Delta f \cdot \Delta r \approx p$. . . . But what is the "political constant" p? I must leave this to a future quantum theory of human affairs. (Quoted in Beller 1998, 29)

There is nothing in the postmodern use of chaos for political ends that can outdo this bit of silliness from the pen of Max Born. And compared to Wolfgang Pauli's speculations on the mind, the worst we could say about Lacan is that he lacked Pauli's vivid imagination. Pauli was especially concerned with "symbols," which he took to involve a "new idea of reality." "The symbol is like a god that exerts an influence on man," he noted. Pauli was in the business of unifying science and religion, "the most important task of our time" (quoted in Beller 1998).

As anyone who has struggled with Heisenberg and especially Bohr knows, these authors can be particularly obscure. Beller's *coup de grace* involves noting what others have failed to note. Wheeler and Zurek's *Quantum Theory and Measurement* is a standard reference work, collecting the most important articles on the foundational problems in quantum mechanics. One of the most famous is Bohr's reply to the Einstein, Podolsky, and Rosen paper. No one noticed that when reprinted in the Wheeler-Zurek anthology, Bohr's article had pages 148 and 149 transposed. After all, the original seemed a word salad, anyway.

What's the moral? I don't know. But I do know what the moral isn't. Mario Bunge says we should "expel the charlatans from the university" (1996, 110). That's just a recipe for disaster. A much better strategy would be to expose the charlatans. Bunge, to his credit, cites quite a few who are not in any sense postmodern. One of those whom Bunge calls a charlatan is Samuel Huntington, the author of a set of "equations" concerning modernization of developing nations that are little more than moonshine apologetics for vicious right-wing social policies. (South Africa under apartheid was a "satisfied" society according to Huntington.)

Postmoderns who are hopelessly confused by quantum mechanics or chaos may be foolish, but do little harm (except, as Sokal fears, divert the Left from useful action). The Huntingtons of the world, however, wield enormous power and influence through their writing and consulting. (Huntington was an advisor to the U.S. government on rural resettlement policy in Vietnam during the war.) Corny as it sounds, the old saw about leading by example comes to mind. Bunge's considerable talents would be better spent effectively exposing the pernicious work of the numerous Huntingtons among us than in trying to "expel" a few postmoderns, who are at worst merely ridiculous.

5
Three Key Terms

A number of terms are bandied about in the science wars, notions that one side too often cavalierly dismisses and the other too often uses to bludgeon opponents. Some of these are pivotal. Getting a grip on *realism, objectivity,* and *values* is essential to making headway in the science wars. There are lots of other important terms, as well, but *rationality* and *relativism,* for instance, can more readily be defined once we have settled on the big three.

Realism

To put it roughly, *realism* is the view that science is more or less successful in telling us how things really are. Put a little less roughly, realism is the philosophical view that says:

1. The aim of science is to give a true (or approximately true) description of reality.

This goal is realizable, because:

2. Scientific theories are either true or false. Their truth (or falsity) is literal, not metaphorical; it does not depend in any way on us, or on how we test those theories, or on the structure of our minds, or on the society within which we live, and so on.
3. It is possible to have evidence for the truth (or falsity) of a theory. (It remains possible, however, that all the evidence supports some theory *T,* yet *T* is false.)

When people say "I'm a realist about X" they usually mean to assert the existence of X. Notice, however, that the characterization of realism given here is at odds with this way of speaking. An atheist, for instance, is a realist. Why? Because an atheist takes the God theory (if I may talk that way) as *aiming* to give a literally true account of things, and of having a *truth-value,* and of there being *evidence* to reasonably settle the matter. However, the evidence, says the atheist, supports the view that the theory is false. Religious fundamentalists agree on all but the final point–they, too, are realists but hold that evidence supports the truth of the theory. In contrast to both the atheist and fundamentalist, a religious *anti-realist* might be someone who holds that religious statements should not be taken literally. The aim of religion on this view is not to give a literally true account of things, but rather to do something else, perhaps to provide moral guidance through storytelling, group ritual activity, and so on. Many so-called liberal theologians are anti-realists of some sort or other.

Realism needn't be all or nothing. There are lots of places where no sensible person would be a realist. First, consider the statement from chess: "Bishops move diagonally." Though true, we cannot be realists about it, since the truth of the statement rests on our stipulating that this is how the game is to be played. It does not comply with the second ingredient of realism, which says the truth is completely independent of us; it's something we discover (if we're lucky), not something we invent. Second, if we ask about sunsets–Are they really lovely?–we're likely to be told that beauty is in the beholder's eye, not part of independent reality. More controversial are examples from ethics and mathematics. Does the truth of "Murder is wrong" and "5+7=12" rest on ourselves in some way? Or do these sentences describe an independent realm of morals and mathematics? Many have found such an independent realm (whether Platonic or God-given) highly implausible and have abandoned any sort of realism in these subject matters. A selective view is possible: One could espouse realism in the natural sciences but not in, say, mathematics or morality.[1]

The easiest way to explain each of the three ingredients in realism is by contrast with some anti-realist account. Recall the instrumentalist view that I mentioned in connection with positivism. For an instrumentalist, the aim of science is not truth, but is instead empirical adequacy. We can accept the reigning theory in particle physics with its quarks, mesons, neutrinos, and so forth, but to accept it is to believe that the theory predicts all observable phenomena to be exactly as we observe them. We should not, says the instrumentalist, believe in addition that there really are quarks that behave as the theory says.

Instrumentalism is one form of anti-realism, a form both popular and influential since Greek times. It's the core of a very long anti-realist tradition in astronomy. From the time of Plato many astronomers have thought the heavens humanly inaccessible. In terms of our three ingredients of realism, they would say something like this: It may be the case that some particular astronomical theory is true (or is false), but its truth doesn't matter, since we have no hope of being able to tell if the theory is true. The heavens are completely out of reach; the meager evidence we have (or could even hope to have) could not possibly decide which theory is true. Consequently, the aim of astronomy should not be the truth, but rather should be mere empirical adequacy. Instead of trying to give a literally true description of the heavens, we should try to tell stories that predict where the observable points of light will be in the sky at various times. If Copernicus' sun-centered system is to replace Ptolemy's earth-centered one, it is only because the Copernican system is better at organizing and predicting when and where the various points of light will be–neither theory should ever be thought literally true.

Copernicus' *De Revolutionibus* was published at the time of his death. His assistant, Osiander, saw it through the press and added a preface (probably contrary to the wishes of Copernicus) stating that there could be no conflict with holy writ, since the aim of Copernicus' work was mere empirical adequacy, not truth. In short, Osiander made Copernicus out to be an anti-realist. Galileo, by contrast, was an out and out realist. When fighting with the Church, he was urged by Cardinal Bellarmine to adopt an anti-realist stance. Galileo could argue that the Copernican system was a better calculating instrument than the Ptolemaic, but, Bellarmine insisted, he should not say that it was true. (Insofar as Ptolemy's theory was thought true, it was not because of scientific evidence, since there could be none that would be decisive, but rather it was considered correct on religious grounds.) Because Galileo refused to adopt this anti-realist line, he was condemned by the Church and lived out his days under house arrest.

There are other ways of being an anti-realist. Let's look at the ingredients of realism in a bit more detail. The first concerns truth and the nature of the connection between theories and the world. Differing views come from many quarters. Realism usually takes this connection to be some sort of correspondence, a relation that exists independently from us. A verificationist, by contrast, claims that the truth of some statement or theory depends entirely on the way we test that theory–truth and evidence are linked. A Kantian, to cite yet another type of anti-realist, claims that many truths stem from the structure of our minds. Causation, for instance, is not part of the objective world, but is in some sense provided by the human mind and imposed on the world. And many instrumentalists think that theories are neither true nor false–they are just systems of symbols that are highly useful in organizing our experience. Each of these is a form of anti-realism and each focuses on rejecting the second ingredient in our characterization of this crucial concept.

Quantum mechanics has provided numerous reasons–all controversial–for being a kind of anti-realist. A typical argument stems from Heisenberg's uncertainty principle. The claim is that some properties, e.g., position and momentum, come in pairs such that a measurement of one leads to its creation and to the nonexistence of the other. Thus, if we were to accurately measure an electron's position, then that electron would have no momentum. This is not the modest claim that we would be ignorant of the electron's momentum once we know the position. Rather it is the much more amazing assertion that the momentum does not exist; even an omniscient god would not know the momentum. Why is this an anti-realist view? It violates the second ingredient concerning independence. We're making the world in some sense through our measuring activities. Realists claim to make *discoveries,* but the uncertainty principle suggests–remember, this is highly contentious–that we somehow *create* the things we describe by choosing to make this rather than that measurement.

Now to the third component of realism, which concerns evidence, namely, that evidence is indeed possible. Skepticism stands in opposition to this by insisting that none of our scientific beliefs could ever be justified. Classical skeptics have no quarrel with the second element of realism; they cheerfully allow that our theories might be true. It's just that we could never hope to know, or even have reasonable grounds for thinking so. It's just dumb luck if we hit on the truth–although we'll never know if we have been lucky or not.

Some anti-realists focus on the notion of independent truth and why it's impossible, others focus on the notion of evidence and why it's hopeless. These are two quite distinct ways of being anti-realists, and they lead to two distinct attitudes toward the *aim* of science. Both say aiming for truth is futile. Instrumentalists adopt the aim of empirical adequacy instead of trying to get a literally true description of the world. Skeptics hold a wide variety of views: some would have us abandon all science as hopeless, others embrace some form of social constructivism (truth is hopeless, so let's try to please ourselves in some other way).

Kantians, verificationists, and others who link truth with evidence still cling to the aim of science as truth. Truth for them, however, is not some literally true description of independent reality; rather it is what the evidence permits us to believe.

Social constructivists typically oppose each of the three features of realism that I listed above. Against the second, they often claim that the notion of truth that realists embrace is incoherent, that it presupposes a nonsensical "God's-eye-view" or some other futile conception. Against the third aspect of realism, constructivists usually claim that the notion of "evidence" is riddled through and through with social content, and is not at all what the typical realist thinks. Moreover, the problem of "underdetermination of theory by data" shows (so the claim goes) that there couldn't be enough evidence to pick out a uniquely correct theory. And against the truth goal of science, many claim that scientists do not aim at the truth, anyway–even if they could–but rather aim at promoting theories that serve their social interests.

Are any of the criticisms of realism fair, whether offered by verificationists, by instrumentalists, or by social constructivists? If called upon to make a dogmatic pronouncement, I'd happily say: The realists are correct, all others are misguided. I dare say no one is asking, but my view is neither here nor there, since we can put the issue of its correctness aside for now. Sorting out the *meaning* of "realism" is our concern, not its truth. In the science wars we too often hear the debate turning on confused discussions of realism vs. anti-realism. This is not where the relevant arguments are to be found. Classic instrumentalists and verificationists are typically opposed to social constructivists, even though they are fellow anti-realists. Why are they opposed? Because they believe that evidence decides which scientific theories should be accepted and which rejected. They take reason and evidence, not social factors, as deciding the matter. Of course, realism stands in striking opposition to social constructivism; but so do a great many anti-realist positions. The real debate turns on something else.

Objectivity

A successful defense of realism would, of course, be a successful attack on social constructivism. Yet, as we just saw, there are several forms of anti-realism that are equally opposed to constructivism. Realism isn't the proper notion to defend here. Instead it is *objectivity.* Realism embraces objectivity, but it is clear, for instance, that instrumentalists and verificationists are also objectivists in a sense that social constructivists are not. In a pinch we might say that scientists are objective in so far as they accept or reject theories on the basis of available evidence rather than on the basis of social or other nonevidential factors.

Let's spell this out with a bit of care. First, we must note that "objective" is usually contrasted with "subjective." There are two distinct senses of this dichotomy, however. One is ontological (having to do with the way things are), the other epistemic (having to do with how we come to know). That gives us four positions in all to worry about. Examples will help us make the appropriate discriminations.

In the ontological sense, a statement is objective if its truth is independent of us. "Water is H_2O," "The racing car is travelling at 200 Km per hour," and "The dog is in the backyard" are plausible examples of objective statements, since their truth (or falsity) is independent of ourselves. (Note the similarity to the second ingredient of realism.) By contrast, "Water is tasteless," "Speed is thrilling," and "Dogs are scary" are all subjective, since their truth depends on us, on our cognitive capacities and background experiences and beliefs. Sugar would have certain molecular properties even if no one existed, but without ourselves to taste it, sugar certainly wouldn't be sweet.

It isn't just statements that can be objective or subjective; we can also classify a property or process on similar grounds: *toxic* is an objective property whereas *tasty* is subjective. Moreover, the distinction comes in degrees. Second, we shouldn't be misled into asking "Tasty to whom? Toxic to whom? Aren't they the same?" "Tasty" and "toxic" may both be relational properties, but one requires the perception of a conscious being to be true whereas the other does not. That's why "tasty" is subjective whereas "toxic" isn't.

Any example that I give of the objective-subjective distinction will, of course, be controverted by a social constructivist. That doesn't matter in the least now. The only point is to explain the distinctions. Once they are clearly understood, they can be affirmed by those who see good reason to do so and denied by those who don't. This goes for both the objective-subjective distinction that I just outlined and the ontological-epistemological distinction, to which I now turn.

In the epistemological sense, a person is objective if that person accepts a statement on the basis of good reasons and evidence. Otherwise, such a person's belief is subjective. Newton was objective in accepting universal gravitation because he had powerful evidence in its favor. People who accept the Genesis account of human origins instead of Darwinian evolution are wildly subjective, since their religious convictions overwhelm all the available evidence.

Once again, these examples will be controversial, but that's not the point. They illustrate the distinctions, which should suit both sides in the science wars. Of course, that doesn't mean that everyone will cite the same cases. No social constructivist would take my examples to illustrate objectivity; they deny any instances of it. The hallmark of social constructivism is subjectivity, in both the ontological and the epistemological sense.

One thing that cannot be overstressed here is *fallibility.* Objectivity does not imply certain truth. Evidence can mislead. The ancients were objective in believing in an earth-centered universe, because the available evidence strongly supported this view. Crackpots who thought otherwise may have been correct, but they have not been vindicated—they were just lucky.

	Objective	*Subjective*
Ontology	Water is H_2O. (A fact about independent nature.)	Water is tasteless. (A fact based on us, not about independent nature.)
Epistemology	Belief that water is H_2O. (Based on standard evidence from chemistry.)	Belief that water is Zeus's urine. (Stemming from smoking too much dope while reading the *Iliad.*)

The interrelations among our four types are interesting. Something could be ontologically subjective, yet epistemically objective. "Vinegar tastes sour" is subjective in the ontological sense, since tastes depend on our cognitive capacities. Nevertheless, our belief that vinegar is indeed sour is based on extensive sampling; we have strong evidence for accepting it. So, in the epistemic sense we're perfectly objective in believing that vinegar tastes sour. When Kuhn claimed that paradigms "create" the world, he, in effect, rejected objectivity in the ontological sense. Whatever properties the world has, it has them by virtue of our imposing a framework or paradigm upon the world. Kuhn cheerfully rejected this form of objectivity, the ontological form. He has also been accused of overthrowing the epistemic sort. But he always denied this charge, the charge of being an epistemic subjectivist. Even though there is no objective world existing independent of us and the paradigms we impose on it, there nevertheless are good objective reasons, Kuhn claims, for choosing one paradigm over others.

It's easy to confuse the ontological and epistemological senses of objectivity and subjectivity. Sometimes they are blurred to humorous effect, as in Mark Twain's quip: "Wagner's music is better than it sounds." It is also easy to be misled by characteristics of objectivity. We think of an objective person as being detached and disinterested. Although "disinterested" and "uninterested" mean quite different things, it's still easy to slide into thinking that someone who is emotionally involved can't be objective. This is nonsense. It is also nonsense to think that someone detached and disinterested is passively uninvolved. Yet this is Barry Barnes's characterization:

> One common conception of knowledge represents it as the product of contemplation. According to this account, knowledge is best achieved by disinterested individuals, passively perceiving some aspect of reality, and generating verbal descriptions to correspond to it. (Barnes 1982, 1)

Social constructivists often complain that their views are being caricatured when criticized, but it's harder to find a more misleading representation of standard philosophy of science than this one. I don't know of a single philosopher who holds anything like this view, which Barnes calls "common." First of all, most philosophers distinguish between mere passive looking and the kinds of observation involved in experimental manipulation–the latter is much more central to science. Even more important, many if not most philosophers of science think we impose frameworks on the world; we try them out, rejecting them when experience suggests a poor fit. And the experience involved in testing the framework's fit inevitably will be (to some extent) theory-laden. This process involves an active mind, not a passive one. It is little wonder that many, on first taking an interest in these issues, find social constructivism appealing when they are presented with a passive or contemplative account of knowledge as the only alternative.

Before leaving the topic of objectivity, I must stress that nothing has been said about how to be objective. Traditionally, being objective usually meant being free of all values. Is this correct? Recent feminists have also upheld the importance of objectivity, but are completely at odds with the claim that this means being value-free. The distinctions between epistemological and ontological forms of objectivity and subjectivity are all-important. It is still an open question how to characterize epistemic objectivity and whether there even is such a thing.

Values

The competing claims that science is "value-driven" and its contrary, that science is "value-free," are both commonplace. Some fear that any value influence on science is corrupting, whereas others merely want the proper values to be at work. But what are values? How do they differ from scientific facts? What connection, if any, might they have to objectivity?

Let's begin with a contrast to facts. There is a longstanding philosophical tradition that distinguishes *is* from *ought.* "Grass is green" and "The proton is heavier than the electron" are *fact*-statements. "You ought to tell the truth" and "You ought to believe the evidence of your senses" are *value*-statements. One easy way to tell the difference between facts and values is to simply look at the corresponding statements. Value statements tend to contain terms like "ought," "should," "must," or "would be better if," and so on. They express norms. The standard view about "is" and "ought" is that they are not inter-derivable. Nothing follows from the fact that grass is green to the effect that it ought to be green. It doesn't follow from the fact that we ought to tell no lies that we don't. Naturally, things are complicated. Let's see how.

The first distinction we need to make is between *cognitive or epistemic values* and other types. Scientific method is full of rules, normative statements, that express epistemic values. Examples include:

- seek empirically adequate theories;
- discard any theory that is logically inconsistent;
- choose theories with the widest scope and explanatory power;
- search for simple theories;

  and so on.

We needn't worry whether these are the right rules. The point is only that they are norms, ought-statements. They express values, not facts. But, of course, these are cognitive or epistemic values. When people claim that science is value-free, they don't mean free of these sorts of values. It is some other type of value that is intended.

The range of nonepistemic values is enormous. Here's a sample:

- Do unto others as you would have them do unto you (the golden rule);
- Keep your elbows off the table (Miss Manners);
- Thou shalt have no other gods before me (First Commandment);
- Live free or die (New Hampshire license plates);
- Always leave plenty of pizza for Jim (my rule);

  and so on.

At first blush, one might think the difference between epistemic (or cognitive) and nonepistemic values simply turns on their connection or lack of connection to science. Not so. It's easy to construct nonepistemic norms that would have a bearing on the practice of science:

- promote theories that will increase the health, happiness, and financial well-being of the author of this book;
- only accept theories that exalt the Aryan race;
- reject any theory that conflicts with the Bible.

These examples seem perfectly silly. When people say science is value-free, it's these sort of values they have in mind. Here are a few that aren't silly in the least, however, but are often vexing:

- inflict no pain on research animals;
- make sure the scientific work environment is friendly to all, especially to women and minorities who have all too often been excluded from the scientific process.

These don't seem to be part of "scientific method," at least not in the narrow sense that concerns the relation of theories to evidence. Someone could, after all, gain a great deal of objective knowledge while treating research animals cruelly, as the history of vivisection illustrates only too well.

Robert Merton (1973) famously proposed four values that he took to characterize the "scientific ethos."

- *Universalism:* the evidence is open to all; there are no privileged observers;
- *Communism:* knowledge is collectively arrived at and is owned by all;
- *Disinterestedness:* we approach nature without prior wishes that it be one way or another; and
- *Organized skepticism:* nothing is immune from doubt.

The rules of scientific method, that is, epistemic or cognitive values, are rules for the acceptance or rejection of theories. People who are secretive, partial, uncritical dogmatists have, alas, often done well using their own rules while violating Merton's. The ability to make sound inferences from data is one sort of virtue, cheerfully sharing data with colleagues is quite another. Presumably we wouldn't want scientists to be free of either of these sorts of values. Those who say science is value-free may prefer to say something like this: "Science exemplifies lots of values, but those values contribute to the growth of objective knowledge. Science is free of values that detract from the truth-seeking goal of science." This formulation will certainly have its shortcomings, but it's a big improvement over the childish drivel about a completely value-free science.

It is often said that so-called scientific creationists (those who reject Darwinian evolution in favor of a special act of creation) let their religious values get in the way of their scientific judgment. Perhaps this charge is justified, but what specifically is involved? The story of the origin of the world and of our own species told in Genesis is a straightforward report of alleged facts; it's not a collection of norms or ought-statements at all. It would be best to say that creation scientists are dogmatically determined to hold on to their religious beliefs that are purportedly factual. So why call them values?

Here's a quick answer. The characteristic thing about fact-statements is that they can be empirically tested, whereas ought-statements cannot. Religious beliefs form a cluster, some of which are norms (Thou shalt not kill) and some purported facts (God created heaven and earth in six days), but none of them are empirically testable. So we lump them all together as values.

The quick answer grounds a distinction that seems evident. If a biologist sets out with a constraint to the effect that any acceptable biological theory must be compatible with Genesis, we would condemn this as value-laden research. But if a chemist sets out with a constraint that any acceptable chemical theory must be compatible with quantum mechanics, we applaud this scientific good sense. The quick answer says that the quantum mechanical constraint was empirically grounded, whereas the religious one was not, and that's the basis of our approval in the one case and disapproval in the other.

I'm sure the quick answer could be justified, but things are inevitably more complicated. During the heyday of fruitful interaction of science and religion, Kepler, Newton, Leibniz, and many others brought religious considerations to bear on scientific theorizing. Kepler geometrized nature on the grounds that God is a great geometer and would want to create the world in the most beautiful of all patterns (recall the example of planetary orbits and the Platonic solids described in Chapter 3). Newton argued that his own account of the workings of nature was best because it required God's occasional intervention to keep the world from running down. Leibniz scoffed at this, saying it showed Newton's physics to be preposterous, since it implied that God was less than a perfect craftsman.

Not only is the notion of God intrinsic in each of these cases, but Kepler, Newton, Leibniz, and others believed they had considerable evidence for the existence of God. Their religious beliefs were not mere acts of faith, but the consequence of rational considerations. After all, prior to Darwin, the argument from design had considerable force. The design argument for God's existence runs as follows: We see design all about us: bees pollinate flowers and flowers feed bees; this could not have come about by mere chance; thus, there must exist a designer, a highly intelligent and powerful creator. Sensible and intelligent people in the seventeenth century quite rightly found the argument from design persuasive. When Kepler and others brought religious considerations to bear on physics and astronomy, they were acting no differently in their day than a contemporary chemist who rightly insists that all theorizing be constrained by what we already believe about quantum mechanics. If belief in quantum mechanics is not a value, then neither was Christian cosmology in the seventeenth century. These were not subjective values that Newton and Leibniz imposed on science, but rather background beliefs that had some degree of plausibility.

But notice the tense: I use *is* when mentioning rationality and values in connection with the constraints of quantum mechanics and *was* when mentioning values and rationality with the constraints of religion. Over time and with new evidence, the rationality of a particular belief changes. It is no longer rational to set religious constraints on theorizing in biology or astronomy or anywhere else. To do otherwise is utterly irrational and very likely dishonest. So we can return to the quick answer: religious beliefs are values in the derivative sense that they are not currently evidentially grounded.

Local vs. Global Constructivism

Scientific orthodoxy (which I outlined in the first chapter) is the view that science aims to tell us how things objectively are. Of course, that's terribly simple-minded. But it's a useful simplification. Social constructivism stands in sharp opposition. Now here's an important qualification that will put the concepts of realism, objectivity and values to work. I'll call the view that there isn't a way anything is independent of our various constructions "global constructivism." And I'll use the term "local constructivism" for the view that orthodoxy is by and large correct, except for certain specific cases. Let's jump immediately to an example.

Research on homosexuality involves an inextricable mix of science and politics. "Pro-gay" and "anti-gay" represent the respective political stances of being in favor or out of favor with improving the social situation of gays and lesbians. Debate tends to flare between the "essentialists" and the "constructivists." (For better or worse, these are the standard terms.) The former claim that one's sexual orientation is a condition that is either biologically determined or imprinted at an early age and remains more or less immutable through life. On this view, one is objectively gay or lesbian, whether one knows and accepts it or not. For a constructivist, on the other hand, one's sexual orientation is fundamentally a *choice,* no doubt a choice conditioned by one's history and upbringing, but a choice nonetheless. Essentialism is compatible with that view of science I have called scientific orthodoxy: sexual orientation is a biological or psychological fact whose nature and causes can be discovered by orthodox science. (This includes both the ontological and epistemological senses of objective, but the emphasis here is on the ontological side.) On a constructivist account, by contrast, such facts simply don't exist, at least not objectively. Instead, one's sexuality is constructed–self-constructed at that.

The essentialist/constructivist debate cuts right across the pro-gay/anti-gay debate. The pro-gay essentialist claims that sexual orientation has a biological character and is perfectly natural, and so completely unobjectionable. It is an immutable characteristic, so it should enjoy the protection of anti-discriminatory laws, just as race and gender typically do. The anti-gay essentialist is likely to see it as a nasty disease that ought to be appropriately treated. By contrast, the pro-gay constructivist sees sexual orientation as a choice, one that should certainly be tolerated, if not celebrated. The anti-gay constructivist agrees it is a choice–a wicked one that must be morally condemned.

The fact that we can find pro- and anti-gay constructivists is a point worth stressing. Social constructivism is not invariably a liberal or left-wing view. Certainly there are many on the Left who hold realist views of science and of knowledge in general. Perhaps in North America a significant number of social constructivists are on the political Left, but the connection is accidental. Some of the currently most prominent and influential social constructivist views stem from Friedrich Nietzsche, a favorite of Hitler, and Martin Heidegger, who joined the Nazi Party.

The truly interesting thing about the sexual orientation controversy is the style of debate. I'll focus on the pro-gay side of things. Even though essentialists and constructivists disagree about the nature and causes of sexual orientation, they share the same political goals, namely to improve the social situation of gays and lesbians. But how do they argue for their respective views?

A typical essentialist is Simon LeVay, who claimed as a result of postmortem studies that a cluster of brain cells in the third interstitial nucleus of the anterior hypothalamus is larger in heterosexuals than in homosexuals. This, he said, shows that "sexual orientation has a biological substrate" (1991, 1034). Another famous work is that by Baily and Pillard on twins. They found that identical twins were more likely to be consonant for homosexuality than were fraternal twins. Given that identicals are closer genetically than fraternals, Baily and Pillard drew the conclusion that "genetic factors are important in determining individual differences in sexual orientation" (1991a, 1093). In a more public piece they stated that "science is rapidly converging on the conclusion that sexual orientation is innate," and they concluded that this is "good news for homosexuals" (1991b, A21).

	Pro-gay	*Anti-gay*
Essentialism	Being gay is biologically fixed. Therefore, it is natural and gays should enjoy legal protection as others do on the basis of gender or race.	Being gay is a disease (possibly genetic) and should be medically treated with the aim of curing the illness.
Constructivism	Being gay is a free choice, one that should be tolerated.	Being gay is a free choice, but a wicked one. It should be condemned like other immoral activities.

Let's review some types of criticisms of these essentialist findings. Some object that LeVay cannot tell which is cause and which effect; perhaps homosexual activity causes differences in the brain. Second, he used routine autopsy reports without questioning the subjects themselves. Many died of AIDS, but could have been, for example, intravenous drug-users. He assumed that subjects who died with AIDS were gay. Two of his subjects he classified as heterosexual, but were known to have engaged in homosexual activity. Perhaps, most important from a constructivist view, by not asking the subjects about their view of themselves, LeVay not only missed a source of evidence, he missed the very thing that is *constitutive* of being gay or lesbian on the constructivist view. Believing one's leg is broken is perhaps evidence that it is, but certainly does not make it so; however, believing that one is gay is intimately related to actually being gay, since being so, on the constructivist view, is fundamentally a choice. Of course, the possibility of self-deception or outright lying clouds this issue, but the fact that self-reports of sexual orientation were not part of LeVay's classification comes close to begging the question in favor of essentialism.

Criticisms of the twins studies include the following: First, Baily and Pillard used a bimodal classification; they assumed that homosexuals and heterosexuals are two distinct populations, when in fact, they may overlap considerably with bisexuals. A continuum of sexual orientations seems called for, but was completely ignored by their study. They used a rather outdated Kinsey test for determining sexual orientation, when more suitable, newer tests were available and should have been used. Furthermore, some of their data are questionable. Often only one of a twin or nontwin sibling participated in the study. The sexual orientation of the other was determined by the declaration of the participant. But this is highly questionable when one considers the reasons for nonparticipation. The participant may not have wanted the other to know of his or her sexual orientation, or the other may have simply refused to participate. Under these sorts of circumstances, how reliable is the participant's assessment of the other's sexual orientation?

Now to the interesting philosophical point. The crucial thing about these criticisms of essentialism that come from constructivists is that they are wholly *within the framework* that I have called scientific orthodoxy. Every objection would be recognized—at least in principle—as perfectly legitimate from the point of view of standard scientific method: Don't use crude tests when better ones are available, don't beg the question when setting up a classification, don't take people's judgments of others at face value, don't ignore obvious alternative possible explanations, don't confuse correlation with causation. The list could go on. These are principles that any champion of orthodox science happily embraces.

When we look at the essentialist vs. constructivist debate over the nature of homosexuality, it's rather easy to see that the sort of constructivism proposed is the local sort, not the global. In every case, the standard techniques of orthodox science are cited and used to undermine particular claims of objectivity (i.e., the claims of essentialism) in favor of claims of subjectivity (i.e., constructivism). Perhaps, one might argue, we can allow the legitimacy of local constructivism in special cases such as this. After all, social science is a dubious business, anyway, isn't it? It turns out that similar forms of reasoning arise even in the heart of theoretical physics.

The picture we often get is one of scientists nailing down the facts, then moving on. This is mistaken in a number of ways. The most common objection focuses on how hard it is to "get the facts." A second objection might go this way: There are no facts to get. Occasionally scientists themselves come to conclusions like this. When they do–though it is invariably contentious–they flag the issue and point out that beliefs are adopted as conventions, not as discovered matters of fact. In connection with this, much has been made (incorrectly) of Gödel's incompleteness theorem (that the unprovable Gödel sentence isn't either true or false, so there is no fact to be discovered) and of Heisenberg's uncertainty principle (if the position of an electron has been measured with complete accuracy, then there isn't any momentum at all). Most popular discussions of Gödel and Heisenberg are misleading and uninformed. (In the Gödel case, the unprovable sentence is true; this seems to follow from the peculiar details of the proof. The Heisenberg case is more difficult to sort out and remains controversial.) Poincaré's view on the geometry of space is a much better example.

According to Henri Poincaré (and later Hans Reichenbach and others), physical space can have some geometrical structure–Euclidean or non-Euclidean–only relative to a particular way of measuring distance.

Before a method of measuring is chosen, there is simply no fact of the matter about the geometry of our space. There are no facts of the matter, says Poincaré, about how measuring rods do or do not change length as they move about in space. Perhaps they shrink or expand from location to location. Here we must choose. If we stipulate that measuring rods do not change, that is, they maintain their lengths as they move around, then we will "discover" that the geometry of our space is, say, non-Euclidean; and if we stipulate that measuring rods change in such and such a fashion, then we might "discover" that space is actually Euclidean. The geometric structure of space is a convention, according to Poincaré, since it hangs on conventional choices we make when we say how measuring rods behave.

The problem is easy to get a grip on when imagining two-dimensional beings who live on the surface of an infinite plane. From our "God's-eye-view" we would say they live on a Euclidean plane. Suppose, however, there was an area on the surface where a bizarre force shrinks measuring rods. (Make this a universal force that similarly affects all materials.) The two-dimensional beings, let us suppose, make the assumption that their measuring rods maintain their length as they move around. Thus, in the area of the bizarre force, they would "discover" as a result of their measurements that space is curved (non-Euclidean) in this region.

Now imagine a second two-dimensional surface, but this time, only finite in extent. This world, let us say from our God's-eye-view, is distinctly non-Euclidean, but there is a force that systematically shrinks measuring rods, so that it would take infinitely many rods of unit length to reach the edge. People on this world would reasonably believe that they live in a Euclidean world.

We can see from examples like this, claims Poincaré, that the geometry which these two-dimensional people "discover" simply results from the conventional choices they make concerning the behavior of measuring rods. Let's remove the God's-eye-view, the higher dimensional space in which the surface is embedded and with respect to which it looks curved. Now there is no fact of the matter about what is really going on independent of what they can determine. The upshot, says Poincaré, is that there is no fact of the matter about the geometry of their space. Once we understand the argument, we can easily see that it applies to ourselves. There is no fact of the matter about our space; whatever we discover will rest on measurement conventions.

Cases like this are highly contentious. Many—perhaps most—scientists and philosophers would reject Poincaré's line of reasoning. I certainly would. But that is not relevant. The moral to be drawn from examples such as this is that when scientists find a case which they themselves believe involves a convention, a choice, or a human artifact, rather than an objective fact of the matter, they flag it. It is singled out for special treatment. It is not passed off as another discovery about independent, objective nature, but is explicitly marked as a human construction.

With these terms—realism, objectivity, value—clarified, we can begin to express the issues a bit more carefully. The debate is not between realists and social constructivists. Rather it is between those who hold that science is (to a significant extent, though perhaps not always) *objective* and those who hold that it is *subjective.* It is the *epistemic* sense of objective and subjective that is central. Sometimes an argument for epistemic subjectivity will appeal to ontological subjectivity, and the argument will be focused there. Yet the real issue is epistemic objectivity vs. epistemic subjectivity. All sides will agree that values play a significant role in science. No sensible debate should be carried on over the claim that science is value-free. Rather, it should be over the status of the particular values involved: Are they grounded in some way? Or do they merely reflect noncognitive interests? As for terminology, "rationalist" and "social constructivist" are convenient abbreviations for the two camps, but they do tend to hide much that is important.

6

The Naturalist Wing of Social Constructivism

In the chapter on the nihilist wing of social constructivism we saw some of the sillier commentaries on science. Sokal and others chewed many of them to bits. They were easy targets, people who for the most part are cultural critics, woefully ignorant of science. In this chapter we look instead at some of the serious practitioners of science studies. Much of this work is performed under the label SSK (sociology of scientific knowledge). Though still a diverse group, they have enough in common to lump them together as the *naturalists,* a term I'll explain below.

A Paradigm Example

To best explain the naturalists, I need to steer a course between two untenable extremes. One is the idea that social constructivism is brand new and came into existence *ex nihilo.* Long before the current crop of social constructivists, there were like-minded others. Perhaps the most brilliant was Karl Marx, who famously remarked that "the ruling ideas are the ideas of the ruling class." Political and economic power, he might have said, brings intellectual dominance. (Note, however, that Marx was also a firm believer in objective science.) The other extreme to be avoided is the never-ending road of precursors. Would we stop with Protagoras, who held that "man is the measure of all things," or ferret out yet earlier versions of anti-objectivist sentiment? I propose starting somewhat arbitrarily with the very influential 1971 article by Paul Forman on the acceptance of quantum mechanics. It's famous, it's controversial, and it's often cited as a model of how to do science studies. For many social constructivists, it's the exemplar, *par excellence.*

How do we explain the rise of the quantum theory in the mid-1920s? In his elaborately titled "Weimar Culture, Causality and Quantum Theory, 1918–1927: Adaptation by German Physicists and Mathematicians to a Hostile Intellectual Environment," Forman offered a sociological explanation. After the Great War, German scientists lost much of their prestige; Spengler had just published his wildly popular *Decline of the West* and Spenglerism was everywhere. The spirit of the times was decidedly mystical and anti-mechanistic. The scientists of the Weimar Republic, says Forman, created noncausal, nondeterministic quantum mechanics to appeal to the German public's mystical and anti-mechanistic outlook, and thereby to regain their high social standing.

By contrast, a more traditional, "rational" explanation might look something like this: The old quantum theory of Bohr and Sommerfeld was not a coherent set of physical principles. The new theory of Heisenberg, Born, Schrödinger, and others (1925–1927) accounted for a wide range of phenomena including the so-called anomalous Zeeman effect, which had been the subject of much perplexity. Consequently, scientists who worked in this field were won over by the explanatory successes of the new mechanics and completely accepted it for that reason.[1]

Where others see rational factors, Forman sees social forces. One need only pay attention to the footnotes of sociological literature in the 1970s and later to see the great importance of Forman's work to the newly emerging style of science studies. To use a Kuhnian expression, it was a new paradigm. The general idea, manifest in Forman's account, is that scientists had social interests and their scientific beliefs are shaped by those interests, not by so-called rational factors. Let's examine Forman's case study in a bit more detail, so that we can clearly see the structure of his argument.

The scientists of the Weimar Republic were living in a hostile intellectual environment, according to Forman. World War I was over and Germany had lost. The public was seriously disillusioned with science and technology. The spirit of the times was mystical and anti-rational. Indeed there was considerable opposition to science, which was seen as mechanical, rationalistic, and linked to causality and determinism. Into this hostile intellectual climate came Oswald Spengler's *Decline of the West,* which claimed that physics expressed the "Faustian" nature of current Western culture. According to Spengler, physics had run its course, exhausting all its possibilities. It stood condemned as a force in opposition to "creativity," "life," and "Destiny." Salvation could only come if science returned to its "spiritual" home.

Several leading Weimar physicists are cited by Forman as stressing the importance of "spiritual values" and acknowledging the "mystery of things." Forman concludes that the concessions were so numerous and extensive that they constituted a "capitulation to Spenglerism" (1971, 55). And so the general "crisis of culture" was embraced by the scientists themselves: "The *possibility* of the crisis of the old quantum theory was dependent upon the physicists' own craving for crises, arising from participation in, and adaptation to, the Weimar intellectual milieu" (1971, 62).

Perhaps the most striking feature of quantum mechanics is the widely accepted belief that it abandons strict causality. Quantum processes have various probabilities of occurring, but they are not invariably determined to do so. (This is one of the features that Einstein so disliked; he claimed that God does not play dice.) Did this new theory that surrendered determinism result from the usual evidential considerations? Not at all, says Forman.

> Suddenly deprived by a change in public values of the approbation and prestige which they had enjoyed before and during World War I, the German physicists were impelled to alter their ideology and even the content of their science in order to recover a favorable public image. In particular, many resolved that one way or another, they must rid themselves of the albatross of causality. (1971, 109)

> [T]he movement to dispense with causality expressed less a research program than a proposal to sacrifice physics, indeed the scientific enterprise, to the *Zeitgeist.* (1971, 113)

Forman's celebrated study became a new model for many historians of science. According to this model we understand events in the history of science not in terms of the empirical evidence, not in terms of theoretical innovations, not in terms of conceptual breakthroughs, but rather in terms of social factors. A group of scientists in Weimar Germany had a social goal–to regain lost prestige. That's why the old quantum theory was rejected and the new quantum mechanics of Heisenberg, Born, and others was adopted.

It's difficult to say why social constructivism has flourished to the extent that it has. One of the reasons is the perceived success of historical case studies such as Forman's. But are they really successful? It's not easy to make a decisive case one way or the other. Certainly, explanations by social factors tend to be more interesting than explanations via dry data and arid inductive inferences. One can read about the events, study the experimental data, and laboriously work through the calculations that lead up to the revolution in quantum mechanics in, say, Jammer's history of the period (Jammer 1966). Of its kind it's a fine work, but it's also hard going. By contrast, Forman's account is a real page-turner, with descriptions of the social atmosphere of postwar Germany, Weimar politics, and so on. Social history is often more fun–but that, of course, doesn't mean that it's right.[2]

Aside from the questionable success of the many case studies, there seem to be two other general considerations that lend sympathy to social constructivism. One of these is politics. Certain political goals are perhaps best served by a constructivist approach to nature. The connection between social issues and constructivism as practiced by the naturalist wing is far from obvious. But for many practitioners, it is certainly present. When people recognize the social nature of knowledge, it is said, they'll adopt a healthier skepticism to the authority of science. The nihilist wing of social constructivists held a similar attitude. We'll look further at this connection in subsequent chapters.

The principal motivation, however, is naturalism, the view that to understand anything–including science itself–is to approach it scientifically. Paradoxically, many see their (science-undermining) social constructivism as the inevitable consequence of a very pro-science attitude.

What Is Naturalism?

There is no crisp answer to this question, but we can at least give some brief idea. Naturalists are motivated by the thought that the natural world is all that there is and the scientific approach is the only way to comprehend it. There is no god, nor any corresponding religious understanding of the world. We may delight in artistic things, but we should not be fooled into thinking there is an aesthetic understanding of the world, an approach which provides knowledge of reality that is different from scientific knowledge. All knowledge is scientific knowledge–there is no other kind. We must, says the naturalist, understand all our activities wholly in natural terms, and this includes the activity of science itself. Neither god nor platonic forms, nor anything else that is in any way mysterious or unnatural plays a part in our genuine knowledge.

Norms appear to figure prominently in our lives, both moral norms and epistemic norms. But where do they come from? What sense can we make of them? How do we know that we ought to refrain from murdering other people? How do we know that when faced with the choice of two theories, we ought to accept the one that is more accurate, more comprehensive, and so on? These are deeply perplexing questions and they cannot be answered in the same straightforward empirical way as we would go about answering such questions as: Is Earth round? Do species evolve? Is there any pizza left?

The naively religious might tell us that moral norms come from God. Clearly this is not a *natural* account of their origin. Plato held that there exists outside of space and time a realm of perfect forms, containing perfect circles, the form of beauty, the pattern of all good acts, the blueprint for all logical reasoning, and much else. Norms, he said, are grounded in this platonic realm. Again, Plato's answer (though it is radically different from the theist's answer) is clearly not a *natural* account of "ought" statements.

Naturalism is a program. Its aim is to eliminate, or reinterpret, or somehow explain away all norms and other apparently unnatural entities. Those naturalists who are biologically inclined try to account for norms as the product of an evolutionary process. Why are we sometimes altruistic? Why are we repelled by incest? Not because these are objectively right or wrong, but because holding such attitudes has had survival value, and so they have become genetically encoded. Those with pro-altruistic and anti-incest attitudes have tended to have more surviving offspring than others, and these attitudes, being biologically stamped upon us, have tended to be passed on to our offspring. This explains away moral norms by a kind of reduction to biological facts.

Similar proposals are offered for epistemic norms. Certain reasoning patterns tend to promote survival; others don't. If Og reasoned: "In the past tigers have regularly eaten people, but I'm sure this one will be quite friendly," then very likely Og is not your ancestor.

Steven Shapin manifests the naturalist attitude well in his discussion of how to do proper history of science. Social constructivists such as Shapin differ from biological naturalists only in seeking social rather than biological causes for belief. The perceived need for such explanations arises in a way perfectly familiar to any working scientist.

> Historical explanation should, and very often does, aim to reduce the domain of the 'coincidental' by searching out links, parallels and connections between one existential factor and another, between existential factors and thought, between one sphere of thought and another. Historians, and not just sociologists, often function as if it is their business to search out such parallels, to cherish them, to attempt to make sense of them when revealed by making every effort to weave them into an integrated narrative. Thus, if it is revealed to us Huttonian geologists, say, tended to be Whigs while Wernerians tended to Toryism, we do not discard this information, we do not cast it adrift in a footnote or an aside; rather we recognize a parallel of this type as the very stuff of history, as a challenge to our capacity for integrative thought. In other words, as historians, we attempt to reduce coincidence in our materials. So one should look for social differences between people maintaining one intellectual viewpoint and another; having found them, we are obliged to make sense of them. (Shapin 1975, 221)

This is just science as usual. Like other naturalists, social constructivists such as Shapin give no credence to religious or platonic or any other sort of nonnatural explanation of science (or, indeed, of anything). Generally, they look to social forces to explain scientific activity, but they would acknowledge that the natural runs well beyond that. In fact, various naturalist approaches are quite compatible. Bloor, for example, would cheerfully accept both biological and sociological accounts of various aspects of science in a complete account of our intellectual life.

Let's now turn to some of the details of his famous "strong programme."

Bloor's Strong Programme

Whenever David Bloor is charged with being anti-science, he is deeply stung. He can't imagine a more unjust accusation. Bloor sees himself as a naturalist who embraces science and its methods completely. He would only add that he has perhaps embraced them more thoroughly and more wholeheartedly than others, and he is willing to accept the consequence of this completely scientific outlook. And what is that consequence? A thoroughgoing social constructivism.

Knowledge and Social Imagery (1976/91) is perhaps the single most important and influential work in the current social constructivist literature. It contains the manifesto of the *Edinburgh School,*[3] known as "the strong programme." What, one might wonder, could the weak programme be? Before Bloor and his like-minded colleagues got to work, traditional sociology of scientific knowledge focused on various issues surrounding science, such as institutions (Who funds them? Why did this one flourish and that one collapse?), scientists (What social class do they come from? Why are there so few women?), relations to governments and corporations (What impact did the Cold War have on science funding? How is the biotech industry influencing research?), and choices of research topics (Why did Galileo take an interest in projectile motion?). But traditional sociology of science would not try to account for the content of any scientific theory. This, according to Bloor, is what makes the traditional approach "weak."

Here's a trite example: Why did Galileo investigate the motion of cannonballs? Why did he believe they moved in parabolas? Weak sociology of science would limit itself to the first question (and perhaps answer that the local military commissioned him to investigate it). They would not attempt to answer the second question, because the answer to that lies in the evidence that Galileo had. Bloor's strong sociology of science tries to answer both questions.

Robert Merton and his school are Bloor's target. Merton's sociology of science does not challenge, but rather complements traditional history and philosophy. Merton, for example, would be happy to account for the growth of science in seventeenth-century England by linking it to Puritanism, as he did in his famous study (1970). Yet not for a moment would he think it appropriate to give a sociological explanation of why Newton's theory of universal gravitation was widely accepted.

Merton formulated a rule of thumb that has come to be known as the *A-rationality principle:* If a rational explanation for a scientific belief is available, that explanation should be accepted; we should only turn to nonrational, sociological, or psychological explanations when rational accounts are unavailable.[4] This is part of the "weak" approach that Bloor explicitly opposes. He insists upon a uniform strategy in dealing with science, one that is utterly thoroughgoing and that penetrates into the very content of scientific theories–in short, he wants a strong programme.

Bloor's motivation, as I said above, is his naturalism, his attachment to science. The idea of naturalism is also popular among philosophers, especially philosophers of science, who hold a variety of versions. The general principle (to repeat what I said above), is this: The natural world is all there is. There are no special methods of investigating things except the fallible methods of empirical science; norms (whether they be the norms of morality or the norms of scientific method) must be explained away or reduced to the concepts and categories of ordinary science, and they must be understood in terms of the natural world.

Naturalism has the advantage of direct and immediate appeal. Want to know about the atom? Study it scientifically! Want to know about disease? Study it scientifically! Want to know about religion? Study it scientifically! Want to know about human society? Study it scientifically! We seem to be tripping right along, and now that we're on a roll, why hesitate? Want to know about science? Study it scientifically! That's what Bloor urges, and it's difficult to object. But what's involved in a scientific study of science itself? Bloor's answer is the four tenets of the strong programme. If you want to adhere to a scientific understanding of science (or a scientific understanding of anything), then these are the main principles with which your account must comply (see Bloor 1976/91, 7).

Causality: A proper account of science would be causal, that is, concerned with the conditions that bring about belief or states of knowledge.

> *Impartiality:* It would be impartial with respect to truth and falsity, rationality or irrationality, success or failure. Both sides of these dichotomies will require explanation.
>
> *Symmetry:* It would be symmetrical in its style of explanation. The same types of cause would explain, say, true and false, [rational and irrational] beliefs.
>
> *Reflexivity:* It would be reflexive. In principle its patterns of explanation would have to be applicable to sociology itself. Like the requirement of symmetry, this is a response to the need to seek for general explanations. It is an obvious requirement of principle, otherwise sociology would be a standing refutation of its own theories.

Two of these principles seem to be perfectly correct–impartiality and reflexivity. The other two either need serious qualification or are simply wrong. I'll postpone examining the problems of the symmetry principle until later in this chapter. And I'll delay serious discussion of the causal principle until the next chapter, where the role of reason is taken up. What is really at issue, as we shall see, is whether reason is a *cause* of belief. For now, let's see what's right about impartiality and reflexivity. We can do this quickly.

Because we're in the business of explaining belief, we're interested in all beliefs, not just the true or rational ones, and not just the false or crazy ones. Optical illusions, for example, are an engaging curiosity and it's nice to have explanations for them, but ordinary veridical perception is also worthy of our intellectual interest. The reigning story of how I manage to correctly see a cup on the desk in front of me is a wonderful achievement of research in physics and physiology. Photons come from the cup and enter my eye, a signal is sent down the visual pathway into the cortex, and so on. Events such as these play a role in explaining how I come to believe that there is a cup on the desk. Whether my perception is veridical or illusory, it needs explaining. The true and the false are on a par. This is Bloor's impartiality principle, and he's perfectly correct to espouse it.

Would anyone think explaining both the true/rational and the false/irrational wasn't the proper thing to do? The impartiality principle hardly seems necessary, yet the A-rationality principle (mentioned above) might be thought to be in conflict. That principle called on giving sociological explanations for beliefs *only* when no rational explanation was available. Actually, there is no conflict between the two principles. Every belief requires explanation, but some will get one type of explanation (say, in terms of social factors), whereas others will get a different account (say, in terms of evidence and reason). This conflicts with Bloor's symmetry principle, but not with impartiality. All sides in this debate can cheerfully embrace the impartiality principle.

What about the principle of reflexivity? Bloor's principle is something that readers immediately pounce on. If all belief is merely the product of various social forces (so this argument goes), then the same can be said of the strong program itself. There can't be any evidence in support of the strong program because Bloor has argued that there is no such thing as genuine evidence. Bloor may well believe the strong program, but that (by his own lights) is because it serves his interests.

This sort of self-refutation problem plagues many views. The skeptic says no belief is justified; thus, the skeptic's own skepticism isn't justified, so we can ignore it. Marx says belief reflects class structure; thus, Marx's own theory merely reflects his social position, so we can ignore it. These kinds of quick rebuttals really won't do, though they are a favorite with beginning philosophy students, as I mentioned in the first chapter. It might be that a particular doctrine is basically correct, but any formulation of it runs into problems. It might be that none of our beliefs is in any way justified, even though saying so runs into paradox.

The reflexivity of the sociology of knowledge is a small problem, perhaps none at all. Yes, says Bloor, social factors cause all belief, and yes, social factors even cause the belief that social factors cause all belief. There is certainly no logical problem here. If there is any sort of difficulty, it stems from thinking that if we know a belief is caused by social factors, then our faith in that belief is undermined. So, if we know that belief in the strong programme itself is caused by social factors, then that belief is also undermined. Bloor simply denies this. He staunchly holds that we can simultaneously hold a belief and hold that the belief is caused by social factors. Perhaps this is implausible (at least for a wide range of cases), but even if Bloor hasn't answered the self-refuting objection, his reflexivity principle certainly defuses it. Tell him that social factors are making him accept the strong programme and he will nod "yes" and smile pleasantly.[5]

Before getting on to the problematic principles of causality and symmetry, let's look at another famous case study. We will see how well it fits the style advocated in the strong programme.

Pasteur, Pouchet, and Spontaneous Generation

In the account of the "spontaneous generation" debate by Farley and Geison (1974), we find a pattern of explanation that is similar to Forman's.[6] Louis Pasteur, according to Farley and Geison, realized that he could better promote his reactionary politics by opposing spontaneous generation. Pasteur was not interested in the scientific truth about spontaneous generation. Rather, he was a right-wing royalist, happy to serve the monarchy and the church, and to oppose the progressive and democratic forces of the day who had allied themselves with Darwinian evolution. A bit more detail will be necessary to clarify the style of their account of this major event in the history of science.

Spontaneous generation is the doctrine that living organisms can arise independently, without parent, from inorganic matter (this process is known as "abiogenesis") or from organic debris ("heterogenesis"). As a serious doctrine, spontaneous generation was more or less dead by the late nineteenth century. Standard histories would give the credit to Pasteur and cite a clever series of experiments that he performed, experiments that demolished the credibility of the theory of spontaneous generation. Farley and Geison reject this account of the historical episode, seeing instead the heavy hand of reactionary politics in Pasteur's activities.

Louis Napoleon came to power in 1848 with the support of the Catholic Church and the most conservative segments of French society. Church and State were allied against two common enemies: atheism and republicanism. And indeed, materialists, positivists, and atheists were often opposed to Church and State alike. For instance, Darwin's *Origin of Species* was translated into French in 1864 by Clémence Royer. Her preface was an explicit attack on the Church, calling it corrupt and blaming it for many of society's ills. Science and politics were intimately and explicitly intertwined. Needless to say, feelings ran high and social tensions were reflected in attitudes to science.

The Pasteur-Pouchet debate began in 1859 when Félix Pouchet published *Hétérogéie, ou traité de la génération spontanée.* Aware of the political climate, Pouchet declared that he was not an atheist and that his version of spontaneous generation was compatible with both religious and political orthodoxy. In particular he noted that the Bible did not rule out God continuing to create after the first six days. In spite of Pouchet's conservative politics, his views on spontaneous generation could be the thin edge of the wedge for those forces that stood against Church and State. Pasteur, according to Farley and Geison, would have none of this.

Louis Pasteur–a deeply conservative, even reactionary man–was happy in the Second Empire. He dedicated one book to the Emperor, another to the Empress, and he greatly benefited from imperial favor. He ran for the Senate as a champion of the established order. He favored the stability of the status quo over free speech, civil liberty, or democracy, which he thought led to "mediocrity."

Pasteur's scientific background is a different matter. He had done a lot of work on fermentation for the wine industry. At the time he argued against a chemical and for a biological account of fermentation, so his later attack on Pouchet seems perfectly compatible with his earlier scientific work, even a natural and reasonable outgrowth. (Traditional accounts of Pasteur stress this point.) However, in Pasteur's work on crystallography, he took the view that molecular asymmetry (which becomes manifest in certain optical phenomena) is intimately connected to life. The forces that were responsible for this asymmetry were ordinary physical forces. This means that abiogenesis could occur under ordinary mechanical circumstances. It is his crystallographic research program that Farley and Geison fix upon. With this in his scientific background, Pasteur should have been more sympathetic to Pouchet's claims. As they remark, "Pasteur could deny the possibility of spontaneous generation only by suppressing part of his own scientific beliefs" (1974, 79).

Why would Pasteur so forcefully oppose Pouchet? As already noted, republicans and atheists had allied themselves with Darwinian evolution. It was widely believed that the evolutionary process needed a start, something like the spontaneous generation of organisms. With this doctrine defeated, some thought, Darwinian evolution couldn't get off the ground. Accordingly, it was his strongly conservative political views that led Pasteur to take the stance that he did against Pouchet. In conclusion, Farley and Geison remark that they

> are persuaded that external factors influenced Pasteur's research and scientific judgement more powerfully than they did the defeated Pouchet. Having formulated his version of spontaneous generation prior to the politically significant Darwinian controversy in France, Pouchet maintained his views with striking consistency in spite of their presumed threat to orthodox religious and political beliefs which he fully shared. By contrast, Pasteur's public posture on the issue seems to reveal a quite high degree of sensitivity to reigning socio-political orthodoxies. (1974, 197)

Let's return now to Bloor's strong programme. How does Farley and Geison's social constructivist account fit in with Bloor's naturalism? That is, how does it satisfy the strong programme?

First, the acceptance (or in this case, the rejection) of a scientific theory, spontaneous generation, is caused by social factors, namely, the desire to uphold a conservative church and political regime. The theory is not evaluated on the basis of what is traditionally known as evidence or good reason. Social factors are the cause of Pasteur's belief. So the causal principle is satisfied. We have found a cause of the rejection of spontaneous generation, and this cause is taken to be a wholly natural entity.

Second, the impartiality condition is satisfied since the rejection of spontaneous generation is explained regardless of the fact that we think it right or wrong to reject that theory.

Third, the symmetry principle is also satisfied. The same style of explanation is given to account for Pasteur's true belief as would have been given if he had adopted the opposite, namely an explanation in terms of his social interests.

Finally, the fourth principle, reflexivity, is not directly at issue, but it is touched on by Farley and Geison, who are clearly sensitive to the matter:

> In writing this paper, we have quite naturally considered the possible influence of external factors on our own interpretation of

> the spontaneous generation debate, and it seems only proper to examine this issue in closing. Especially because we find distasteful many of Pasteur's religious, political and personal attitudes, our interpretation may well differ from that of historians and scientists of more conservative persuasion. (1974, 198)

The Symmetry Principle

The symmetry principle carries the load for Bloor's theory, as it does for any social constructivism that has a naturalistic orientation. Bloor wants the same types of explanation for various phenomena regardless of our evaluation of those phenomena. Impartiality demanded an explanation regardless of the rationality of the belief. That's a principle we can all accept. Symmetry demands more, however; it requires the same type of explanation of a belief regardless of its rationality or irrationality.

The A-rationality principle might well be called the anti-symmetry principle, since it demands a completely different response. Recall what it says: give rational explanations to rational beliefs, and look to social (or psychological or other natural) factors only when a belief is irrational. It's worth reminding ourselves just how common-sensical the A-rationality principle really is. Suppose George believes that Martha has the flu. If asked why he believes this, we might answer that George took her temperature, which he noted was high, saw her vomiting, and knew of other flu patients with whom Martha had been in recent contact. That's how we'd explain George's belief; we cite the very evidence that George himself has for his belief. But what if John believes that far from having the flu, Martha is actually reacting to her recent abduction by aliens with whom she was forced to have sex and by whom she might be pregnant? How do we explain John's belief? John might say that he saw the departing spaceship, that he believes that this sort of thing happens often, that Martha protested too much in denying it, and so on. Most of us give this no credence at all. We explain John's belief in terms of his unstable mental state, his misperception of a weather balloon, and so on. We certainly don't try to explain his belief in terms of the evidence that he cites. In short, we treat George's rational belief one way and John's irrational belief another. We do not treat them symmetrically, as Bloor would like us to do.

In justifying the symmetry principle, Bloor remarks: "the sociologist seeks theories which explain the beliefs which are in fact found, regardless of how the investigator evaluates them" (1976, 3). This claim is at once deeply right and deeply wrong. First, let's examine the sense in which it is right.

In discussing these issues Bloor often runs truth, rationality, and success together. Given his general outlook, this may be justified, but *prima facie* these are quite different concepts. When explaining belief, *truth* is hardly ever an issue. Why? Because truth is not transparent. If it were, we would have no problems doing science and there would be nothing over which to have philosophical debates. Bloor is quite right: we should explain, say, Ptolemy's belief that the earth is at the center of the universe, ignoring the fact that we think his belief is false. Truth is one thing, rationality is quite another. Although rationality is not perfectly transparent, it is more or less so. We can follow a chain of inferences and evaluate it in a way that we could never determine the truths of astronomy. The way we normally and properly explain a belief is in terms of the evidence available to the historical actors–not the evidence available to us. We evaluate their reasoning processes, not the factual correctness of their assumptions or their conclusions. So, when Bloor says *our* evaluation of the belief should play no role in our explanation of it, he's correct if he means our evaluation of its truth. But he's dead wrong if he means our evaluation of its rationality.

If Bloor confined his symmetry principle to the limited requirement that the same types of explanation should be given for true and for false beliefs, then he would have our cheerful acquiescence. He has no grounds whatsoever to extend it to requiring the same types of explanation for rational and for irrational beliefs. Until he can provide such grounds, there is no reason to accept his symmetry principle, a principle so contrary to our standard practice in science and in daily life.

We might be tempted to dismiss the symmetry principle out of hand. But that would be a mistake, if only because so many take it seriously. Why do people find it appealing? Bloor seduces us into acquiescence when he remarks:

> The aim of physiology is to explain the organism in health and disease; the aim of mechanics is to understand machines which work and machines which fail; bridges which stand as well as those which fall. Similarly the sociologist seeks theories which explain the beliefs which are in fact found, regardless of how the investigators evaluate them. (1976/91, 5)

It sounds so reasonable, how could we object? It is only after considerable probing that we begin to see how untenable the symmetry principle is. Let's begin by asking what Bloor means by demanding the "same type of cause" to explain rational beliefs as to explain irrational beliefs. Obviously, it can't mean identical cause. Then we would explain standing and fallen bridges citing "the same type of cause," namely, let us say, an earthquake. This, of course, is absurd, since earthquakes do not explain why a bridge is standing. We could try understanding the demand for the "same type of cause" as simply a demand for a causal explanation (in accord with the causal principle). But this would be too liberal, since even rationalists would happily agree to this understanding of the symmetry principle, once it is accepted that reasons can be causes. Then the habit of explaining rational beliefs in terms of evidence and irrational beliefs in terms of social factors would be in happy accord with the symmetry principle after all. Obviously, this won't do either. Bloor clearly has something else in mind. I suspect that something like the following will be more congenial to him.

We turn to physics to explain some phenomena, to physiology to explain other phenomena, and so on. Let us take the symmetry principle to be demanding only this: If we turn to physics to explain phenomena *P*, then we should also turn to physics to explain any other phenomena *P′* that is appropriately related. Standing and fallen bridges are appropriately related. We needn't use the same principles of physics in the two cases, but we must draw on physics in each. Similarly, if we turn to physiology to explain sickness, then it is to physiology we should also turn in order to explain good health. If we can use sociology to explain irrational beliefs, then we must use sociology to explain rational beliefs, too. Again, it needn't be the same sociological factors that explain both, but when we call on sociology in general to account for one, then we must do the same for the other.

This sounds quite plausible, until we ask about the boundaries of physics. What counts as physics and what does not? Disciplinary boundaries are somewhat arbitrary things. If I explain the standing bridge in terms of physics and the fallen bridge in terms of chemistry (properties of the cement foundations that cause it to give way under gravity), am I really departing from the symmetry principle? We can easily imagine someone claiming (and others disputing) that chemistry is really just physics. This is perhaps a fly in the ointment, but not a serious difficulty. There are genuine difficulties for the symmetry principle that are quite serious, however. Here's one.

When we explain anything we usually draw on not just a single, isolated theory, but on various other background assumptions, including other theories. For example, in explaining the observed motion of Mars, we appeal to Newton's three laws of mechanics, Newton's theory of universal gravitation, assumptions about where Mars is at some initial time, and some theory of optics that tells us how telescopes work and how light behaves in the atmosphere. From these premises we are able to explain the subsequent observed motion of Mars.

In this explanation, the focus of our attention might be Newton's theory of universal gravitation. But from a logical point of view, the optical theory is just as responsible as universal gravitation for explaining what we observe. Does this mean that whenever we are trying to explain any sort of planetary phenomena, the same optical theory must be used? What if we are observing from a satellite, outside the atmosphere? What about using radio astronomy techniques? These would seem to be prohibited by the demands of the symmetry principle. If we used that optical theory once for Mars (and we haven't disavowed its use in that instance), we have to use it elsewhere. This is bordering on the absurd, and we can push it right into absurdity by considering the next example.

Consider the style of Forman's explanation for the Weimar scientists' acceptance of quantum theory. It has this form:

1. The Weimar scientists had lost their high social standing and they wanted to regain it.
2. They realized that if they accepted nondeterministic quantum mechanics, then they would likely rise in public esteem.

∴ They adopted quantum mechanics.

Bloor would accept this as an explanation of belief in terms of social causes. If we are prepared to use social factors to explain beliefs that are irrational, then we must also use social factors to explain beliefs that are judged to be rational.

Notice the crucial second premise, however. This belief is the result of reason and evidence. The Weimar scientists determined–rationally on the basis of the evidence they had–that a certain course of action would best achieve their goal. Notice that they did not come to believe that they could regain their lost prestige by forming a football team or by taking up ballet. This means that reason and evidence play a role, even when social factors are used to explain a belief. Reason is as much the cause of their belief as are social causes. Now, by the symmetry principle, Bloor is stuck having to accept the causal role of reason in explaining all beliefs. Not even I believe that he should do that. Suppose the Weimar scientists had come to believe that they could achieve their goal of regaining lost prestige by collectively mooning the public from the top of the Brandenburg Gate. Clearly, sometimes reason and evidence play a causal role in the explanation of belief, and sometimes not. Sometimes social factors play a causal role in the explanation of belief, and sometime not. And that's the end of the symmetry principle.

There is another response to Bloor's symmetry principle that a rationalist could make–accept it. If we point out that some explanations use reason and evidence, we can insist (by symmetry) that all should do so. This is not as far-fetched as it may seem. We could say everyone is rational all the time. The crucial difference is in the goals sought. We can think of rationality as a means to an end, and in science we usually take the end to be the truth. But if the end is something else–regaining lost prestige or protecting the Second Empire–then we can see the actions of the Weimar scientists and Pasteur as rational, after all. They merely had social goals, rather than truth, as their aim. The symmetry principle will be satisfied–reason and evidence are the causes of belief in every case. Only the goals have changed. Needless to say, this understanding of the symmetry principle is no help to social constructivists.

Latour on Symmetry

In spite of his highly touted symmetry principle, there is a glaring asymmetry in Bloor's approach to science. To make this clear, let me begin with a short analogy. Newton held that space is an absolute entity, existing in its own right, completely unmovable. This absolute space is the origin of inertia in material bodies. Thanks to space, bodies (with no force on them) move at a constant velocity in a straight line. Of course, many objected to the very idea of absolute space, but there was also a second problem that troubled many physicists up to the time of Einstein. The problem was the glaring asymmetry between space and matter: space had an effect on matter (making it move in straight lines), but matter had no effect on space. This deep conceptual problem was resolved in Einstein's general relativity: space tells matter how to move and matter tells space how to bend. The old asymmetry between space and matter in Newtonian physics was destroyed and a symmetry between them introduced.

There is a similar disturbing asymmetry in Bloor's approach: causation goes only one direction. Social factors cause scientific belief, but it's a one-way street. Belief seems to be a kind of epiphenomenon, the result of natural forces, but without causal influence back upon them. This is a serious conceptual problem in Bloor's program, every bit as serious as the asymmetry involved in Newton's unmovable space and every bit as problematic.

This is one symmetry-related problem in Bloor's view of the sociology of scientific knowledge, but there are others. Bruno Latour and some of his colleagues have rejected Bloor's version and in its place proposed a much more radical and extensive symmetry principle. What they want to overcome is the asymmetry between the natural world and the social world, between nature and society. By now it should be clear that social constructivism is very far from being a monolithic bloc. The difference between the nihilist and naturalist wings, as I've called them, is glaringly obvious, but there are major differences inside the naturalist camp, as well. The battle over the symmetry principle marks the most significant split.

In *The Pasteurization of France,* Latour puts humans and microbes on a par. A typical sociological account would look for various social factors that led Pasteur and others to believe that there are microbes, that they make people sick, and that they can be controlled by what is now called pasteurization. Such an account would give primacy in the explanation to the social realm; the so-called facts about nature are merely what results from a process of social negotiation (the facts are a social construction). In contrast, Latour gave equal billing to nature; the microbes themselves are part of the plot. Just like humans, microbes have interests, says Latour, and will decide to cooperate or not in the eventual outcome.

Michel Callon (a close colleague of Latour) endorsed this radical symmetry in his study of scallops and scallop cultivation.

> The observer must abandon all a priori distinctions between natural and social events . . . The capacity of certain actors to get other actors whether they be human beings, institutions or natural entities, to comply with them depends upon a complex web of interrelations in which Society and Nature are intertwined . . . If the scallops are to be enrolled, they must first be willing to anchor themselves to the collectors. But this anchorage is not easy to achieve. In fact the three researchers will have to lead their longest and most difficult negotiations with the scallops. (1986, 201–211)

At first blush Latour's extreme symmetry principle seems a genuine improvement: The structure of society can be the result of science just as science can be the result of social input. But the principle in action is completely bizarre. Not only do scallops negotiate with fisheries researchers, but, presumably, protons have negotiated their cloud chamber activities with physicists, and tulips bloom in the spring thanks to a deal they worked out with our botanists. Of course, this is completely preposterous and one can only marvel at the lengths to which Latour and Callon will go to banish any asymmetry between the natural world and the social.

Quite aside from the intrinsic absurdity of Latour's radical symmetry principle, it still fails, even in its own terms. He has destroyed the asymmetry between the natural and the social in one sense, but not in another. His extreme symmetry is successful in putting microbes (nature) on a par with human scientists (society). Yet notice the character of the interaction between nature and society. Microbes interact with humans, but they do not do this biologically, or chemically, or gravitationally—*they interact socially.* Far from completely eliminating the natural-social asymmetry, Latour has simply endowed nature with various social interests. Nature has become a pack of rival politicians, and science studies is merely the reporting of how our diplomats–the ones we call scientists–cut a deal with them. There is no more a role for an independent nature in Latour's version of the symmetry principle than there is in Bloor's.

Schizophrenic Method

Readers of social constructivist literature, even sympathetic readers, feel an acute sense of discomfort when they notice that constructivism seems to apply only to the scientists under study, not to the sociologists who are doing the study. Thus, Pasteur and the Weimar physicists do not have a grip on nature, but are constructing their theories under the influence of social forces. By contrast, Farley and Geison and Forman are describing the objective facts about those scientists and the social forces acting on them. There are no facts about the quantum world or about spontaneous generation, but there are indeed facts about the influence of Spengler on the German public and facts about Pasteur's reactionary politics. The tension is striking, and those who don't feel it simply haven't noticed a glaring problem. Harry Collins and Steven Yearley have noticed it, and they are prepared to bite the bullet.

> Natural scientists, working at the bench, should be naive realists–that is what will get the work done. Sociologists, historians, scientists away from the bench, and the rest of the general public should be social realists. Social realists must experience the social world in a naive way, as the day to day foundation of reality (as natural scientists naively experience the natural world). That is the way to understand the relationship between science and the rest of our cultural activities. (1992, 308)

The position they adopt is quite unsettling, even to those sympathetic to social constructivism. This is what Latour objects to when he demands a stronger form of symmetry. There are several things that should be mentioned about this somewhat schizophrenic view.

First, it denies the reflexivity principle, as Bloor propounded it. Bloor's principle says that whatever the sociologist of scientific knowledge says about science applies in principle to the sociologist as well. Bloor is perfectly clear about this and accepts the consequence that various social forces are making him adopt this view. Collins and Yearley reject the point. They say in effect (to use Bloor's naturalistic language): What holds for science does *not* hold for the science of science. Bloor justifies his constructivism on the grounds that it is scientific, and he sees reflexivity as part of that justification. By adopting their two-tiered view, Collins and Yearley have abandoned a key part of the naturalistic motivation for this whole approach to understanding science. I needn't dwell on how *ad hoc* this is.

Second, the level at which sociologists and historians work is not going to be stable. The social realism of this realm will often give way on examination in particular cases. A famous example will nicely illustrate. Boris Hessen was a Soviet scientist who published a Marxist study, *The Social and Economic Roots of Newton's* Principia (1931/71). To an earlier generation of sociologists of knowledge, Hessen's work was very influential. More recently Loren Graham, in a study of Hessen, has claimed that when Hessen wrote his study, both he and Soviet physics itself were under fire, and that Hessen was trying to protect Einstein's relativity from attacks by misguided materialists. "Hessen's paper on Newton was carefully crafted to support this defensive effort and simultaneously was aimed at strengthening Hessen's own political situation" (Graham 1985, 705).

Using the Collins and Yearley formula, we seem to have the following:[7]

1. In Newton's eyes Newton is *describing* the natural realm.
2. In Hessen's eyes Newton is *constructing* the natural realm and Hessen is *describing* the social realm of Newton.
3. In Graham's eyes Hessen is *constructing* the social realm of Newton.

There is a real clash here between (2) and (3). Indeed, every analysis of an episode in the history of science can be followed by a meta-analysis, and that in turn by a meta-meta-analysis, and so on. And each time one is completed, the lower analysis changes from realist description to social construction. This is a bizarre form of relativism–quite a new kind–and perhaps even more hopeless than the old kind, since it requires a single reader to switch views schizophrenically in the midst of a single topic. Ordinary relativism says that rival theories are *both* true, but it does not ask any individual to actually believe both. But if I'm studying Newton and am led to Hessen's and then later to Graham's accounts of the matter, I am obliged by the Collins and Yearley proposal to pull off a mental stunt that I suspect is psychologically impossible, not to mention logically out of the question.

Anthropology in the Lab

Anthropology is a science. Some might dispute this, but let's grant it for now. Being a science, it would seem to satisfy the tenets of the strong programme; providing an anthropological account of science is another way of being a naturalist. And why not? Anthropologists regularly study this or that exotic society, so why not use the same techniques to study the peculiar society of scientists? This is the attitude of Latour and his co-author Steven Woolgar, who remark, "Whereas we have a fairly detailed knowledge of the myths and circumcision rituals of exotic tribes, we remain relatively ignorant of the details of equivalent activity among tribes of scientists" (1979, 17).

The story told in their now classic work, *Laboratory Life*, is an amazing one about the creation/discovery of TRH (for thyrotropin releasing hormone). This is a very rare substance produced by the hypothalamus, which plays a major role in the endocrine system. TRH triggers the release of the hormone thyrotropin by the pituitary gland; this hormone in turn governs the thyroid gland, which controls growth, maturation, and metabolism. TRH was discovered (or constructed) by Andrew Schally and Roger Guillemin, independently, and they shared the Nobel Prize in 1977 (as co-discoverers, though each disputed the other's claim). The amount of physical labor involved in isolating TRH is mind-boggling. Guillemin, for example, had five hundred tons of pig's brains shipped to his lab in Texas while Schally, in San Diego, worked with a comparable amount of sheep's brains. Yet, for all their labor, the quantity of TRH extracted was tiny.

An identification problem stems from the lack of any significant amount of the hormone. As the existence of the stuff is questionable, any test for its presence is problematic. Latour's philosophical claims about facts largely turn on this. Consider gold: we have lots of this stuff; it's observable, easily recognized by ordinary people. We have paradigm samples. To protect ourselves from "fool's gold" and fraud, tests (assays) have been developed. How do we know that a particular assay is a good test? We simply use standard samples of gold and samples of nongold, and consider an assay to be a good one insofar as it can reliably distinguish between them.

Such a procedure is not possible in the TRH case, however. We don't have independent samples that we can use to "test the test." Different bioassays for TRH were developed by different research teams, but without a standard sample of TRH there is no independent check on any of these proposed bioassays. Thus, there is no way to be sure a bioassay is "true to the facts," as Latour and Woolgar put it. The fact at issue is this: There is a substance in the hypothalamus that releases the hormone thyrotropin from the pituitary and its chemical structure is pyroGlu-His-Pro-NH_2.

The *existence of the fact* rests on the *acceptance of some particular bioassay;* they stand or fall together. At least this is what Latour and Woolgar argue in *Laboratory Life.* The exact claim is this: "Without a bioassay a substance could not be said to exist" (1979, 64). Perhaps this is obvious to "anthropologists in the lab," but no argument is offered for this crucial claim. It is also taken as obvious that since there is no *direct* test of the bioassay, it must have been adopted as a result of *social negotiation.*

In outline the case Latour and Woolgar make runs as follows:

1. TRH exists if and only if bioassay B is accepted.
2. There is no direct test for B.;

∴ B is accepted as a result of social negotiation.;
∴ TRH is not discovered; it is a social construction.

It's an interesting story with a plausible conclusion. On reflection, however, we can see that neither premise is acceptable. For example, the first premise implies that there was no gold until there was an assay for it. This is an incredible piece of idealism (i.e., believing makes it so). There is, of course, a long tradition of theories of meaning and truth that link facts to tests (e.g., verificationism). Yet a plausible version of such a theory must be able to distinguish the truth from what is tentatively believed to be the truth. Often this is accomplished by defining the truth as what is "verifiable under ideal conditions" or "verifiable in the limit." The link between facts and tests suggested in *Laboratory Life* is much too crude to be in any way believable. Perhaps Latour and Woolgar have something tamer in mind. Their exact claim, once again, is: "Without a bioassay a substance could not be said to exist." Perhaps they mean that without a test we have no grounds for *asserting* the existence of the stuff. This is perfect good sense, something with which we can all agree, since it leaves open whether the stuff in question (gold, TRH) actually exists. Their existence is *independent* of any test. Consequently, facts are not social constructions after all, contrary to the tale they want to tell.

What about the second premise of the argument? The assumptions involved in this are reminiscent of Quine's "web of belief." Propositions are connected to one another in a network; the truth of any proposition is connected to every other proposition. Some of the connections are strong, others are weak; but in any case the network is huge. The contrast is with some sort of atomistic view of meaning and truth: each proposition has its meaning and its truth-value independent of every other proposition. The upshot of an atomistic view is that propositions can be tested for their truth-value in isolation, whereas on a network view of meaning it is the whole network that is put to the test. The picture painted by Latour and Woolgar is again initially plausible, since much of our belief about any substance is intimately connected to whatever bioassays we have adopted, but then their argument sadly breaks down in simplemindedness.

Sociologists of science are often acutely sensitive to this network feature of scientific beliefs and very skilled at making the multitude of complexities in our web of belief manifest. Philosophers of science, by contrast, are often guilty of ignoring these subtleties, and would do well to follow the example of sociologists and historians. However, when it comes to drawing morals from what we know about the network of belief, many sociologists of science become suddenly and strangely simplistic. According to Latour and Woolgar, for instance, the network consists of only two propositions: "This is TRH" and "This is the bioassay that detects TRH." They are linked in such a way that they survive or die together.

Harry Collins is another anthropologist in the lab and has produced a large number of impressive and influential case studies. He, too, likes the network analysis of belief and uses it to great effect in his study of

Weber's detection of gravity waves (Collins 1985). Weber built a large complicated apparatus to detect gravity waves and claimed success in actually detecting them. Others were highly skeptical. Collins made this claim: The pair of propositions "There are gravity waves" and "This is a working gravity wave detector" stand or fall together. To accept (or reject) one, he said, is to accept (or reject) the other. The idea behind it all for Collins is the same as it is for Latour and Woolgar. They accept the analysis that meaning and truth are dependent on a web of belief, but the actual network they focus on consists of a mere pair of propositions.

Such simplemindedness is easily countered. Consider some of the reasoning that went into choosing the bioassay which has been adopted for detecting TRH: Rats are used instead of mice because mice are believed to have more sensitive thyroids; males are used instead of females because the female reproductive cycle might interfere; eighty-day-old rats are used since that is the age when the thyrotropin content of the pituitary is greatest; and so on. Of course, these are all fallible considerations, often loaded with unwarranted assumptions. Some of these assumptions clearly involve social factors. For example, why take females at certain stages of their reproductive cycle as somehow abnormal? Nevertheless, the crucial thing is that they are *independent* reasons for thinking that the particular bioassay adopted is the correct one for detecting TRH. (I don't say independent of all social forces, just independent of the researchers themselves.) This mass of complications is a multitude of connections into our overall network. *Contra* Latour and Woolgar (and the same goes for Collins), we do not have a little circle consisting of only two propositions that will stand or fall together–we have a very much larger network. The bioassay is supported by numerous far-reaching strands.

It may be a social construction that rats have more sensitive thyroids than mice, but it was not constructed by the TRH scientists. For them, it functions as a kind of external constraint; it is a given that they have to work with. Consequently, the claim that the bioassay was accepted through their social negotiations will be much harder to sustain. The sociological story told by Latour and Woolgar, and by others such as Collins, seems plausible only because the networks they posit are so small. It's easy to imagine a group of scientists negotiating over a pair of propositions, a pair that together sink or swim, but when they have to work that pair into the rest of a huge network over which they have little or no control, the sociological case is much less plausible. Once we allow that people have to work (even when they have an eye on their own interests) with facts that are *not constructed by them,* it's only a small step to admitting the possibility that sometimes people have to work with facts that are *not constructed, in any sense.*

A Word About Methodological Holism vs. Individualism

Emile Durkheim introduced the notion of *social cohesion* to explain certain aspects of suicide. (Catholic societies had lower rates than Protestant societies. Durkheim explained this by saying they had greater social cohesion.) Social cohesion is a strange notion and Durkheim realized it. He claimed that the greater social cohesion of Catholic societies is a social fact, a fact about a social group that is not reducible to facts about individuals. This started a major debate in the philosophy of the social sciences. In opposition to this holism is the view called methodological individualism. The idea is that claims about social groups are really just shorthand for (or somehow reducible to) claims about individuals. If I say the nation wants a new government, I don't really mean to ascribe a wish to the nation. Rather I mean that Joe wants a new government, and Mary wants a new government, and so on.[8]

Naturally enough, in discussions about the sociology of scientific knowledge, questions about methodological holism vs. individualism arise. It's fair to say that in a full account, this debate will have to be settled one way or the other. However, it is also fair to say that for our purposes we can avoid the holism vs. individualism issue. For example, favoring methodological individualism (as I am inclined to do) has no bearing on the correctness of, say, Forman's account of the Weimar scientists. A rationalist says "the Weimar community of scientists accepted quantum theory on the basis of the evidence" and means that individual scientists accepted the theory on the basis of the evidence. A sociologist of knowledge says "the Weimar community of scientists accepted quantum theory because of their interests" and means that individual scientists accepted the theory because of their individual interests. On the other hand, if we adopted the outlook of methodological holism, we would talk of the whole community accepting a theory on the basis of the evidence or because of the whole community's interests. Either way, we still have the debate between the rationalist and the sociologist. The dispute between the methodological holist and the methodological individualist does not determine the outcome of the rationalist-sociologist debate.

Does Social Constructivism Undermine Science?

Social constructivists often say that their approach does not undermine science, but only a false understanding of it–the one typically provided by philosophers. Indeed, as Bloor, Barnes, and Henry in their recent book *Scientific Knowledge: A Sociological Analysis* put it, they "honour science by imitation" (Bloor et al. 1996, viii), and they see themselves as merely providing a scientific understanding of science itself. Their account of science may be correct, but the claim that it doesn't undermine science is baffling. This is a quite different point than the one above about the strong programme undermining itself. It's worth a moment's consideration.

There are special cases in which claims about social factors seem perfectly plausible. Money, for example, is a social institution. A coin has value only insofar as people agree that it has. When we come to understand the social basis of money, it is not suddenly deprived of its value. This makes perfect sense, but science is quite disanalogous, since there is nothing corresponding to an explicit contractual agreement about the facts of nature in the same way there is such agreement about money.

The second point ("we honour science by imitation") is of very limited effect in upholding the status of science. Bertrand Russell once said, "Naive realism [i.e., common sense] leads to physics, and physics, if true, shows that naive realism is false. Therefore naive realism, if true, is false; therefore it is false" (Russell 1940, 15). Of course, he put it as dramatically and paradoxically as he could, but the point is a simple one and has the logical form of the tautology: $(P \rightarrow \sim P) \rightarrow \sim P$ (in English: If proposition P implies not-P, then not-P). Though odd-sounding, it is perfectly correct. Indeed, no matter what the component proposition P may be and no matter whether P is true or is false, the whole complex proposition must always be true.

The naturalism of Bloor et al. (and numerous other writings) has a similar form that runs like this: (1) We cheerfully embrace the wonderful methods and results of science and uphold its supreme authority. (This is P.) (2) We then apply these methods to science itself, and we discover that the methods and results of science are not objective, after all, but are rather social constructions. That is, we find that there is nothing special about science; it's just another social institution clamoring for power. (This, in effect, is $\sim P$; and we have shown that $P \rightarrow \sim P$.) From the fact that $P \rightarrow \sim P$, it follows immediately by elementary logic that $\sim P$.

At this point, however, and contrary to all sound reasoning, Bloor and his colleagues in effect say: "We're naturalists and we love science. We're not undermining it; just look at (1) (i.e., we gladly accepted P)." Yet this ignores the whole line of reasoning. Instead, they should draw the conclusion $\sim P$, thus rejecting the authority of science.

Many of the classic case studies have this effect. Consider, yet again, Farley and Geison's account of the Pasteur-Pouchet debate on spontaneous generation. Would anyone still be inclined to accept Pasteur's position *after* having read Farley and Geison who point out that it was Pasteur's reactionary politics and not ordinary evidential considerations that were at work? Supposing Farley and Geison to be correct, then if we knew *then* (i.e., in the late nineteenth century) what social constructivist historians claim to know *now*, our beliefs *then* would have been seriously undermined. To think otherwise is to fail to grasp the very significant implications of serious social constructivist analyses of historical events.

This in itself is not to say that constructivist analyses are either right or wrong. If they are right, however, the implications for our understanding of science are very significant and very different from what Bloor claims. Don't propose a revolution, then deny the revolution will change anything.

7
The Role of Reason

Kuhn

Thomas Kuhn's *Structure of Scientific Revolutions* (1962/70) did much to stimulate and promote sociological approaches to science, something I've stressed repeatedly. As characterized by Kuhn, the switch from one paradigm to another does not seem to rest on evidence and reason as these notions are normally conceived; so the explanation of a shift in belief must appeal to some other type of cause, usually some sort of social factor. Undeniably, there is much in Kuhn's book for the would-be social constructivist. For instance, no constructivist would be displeased with Kuhn's remark, "Like the choice between competing political institutions, that between competing paradigms proves to be a choice between incompatible modes of community life"(1962/70, 94). This smacks of the social. Kuhn offers an argument suggesting that it must be so: "As in political revolutions, so in paradigm choice–there is no standard higher than the assent of the relevant community" (ibid.). It is not reason and evidence that play the decisive role, says Kuhn: "In these matters neither proof nor error is at issue. The transfer of allegiance from paradigm to paradigm is a conversion experience that cannot be forced" (ibid., 151).

These passages, as I said above, strike most readers as saying it is not reason and evidence, but rational factors that determine scientific decision-making, at least when competing paradigms are at issue. The evaluations of Kuhn's views by philosophers were not favorable: a theory of "mob psychology," said one (Lakatos 1970, 178); another complained that paradigm change "cannot be based on good reasons" (Shapere 1966, 67).

On the other hand, social constructivists are not in the least upset with the demise of reason, about which they were skeptical, anyway. Once again, here is Bruno Latour: "'Reason' is applied to the work of allocating agreement and disagreement between words. It is a matter of taste and feeling, know-how and connoisseurship, class and status. We insult, pout, clench our fists, enthuse, spit, sigh, and dream. Who reasons?" (1988, 179f).

Kuhn is often cited in support of social constructivism. But what, exactly, is his view of the role of reason in explanation? Barry Barnes speaks for many social constructivists when he gives this answer:

> That judgements made in the course of normal science are culturally specific and conventionally based is clear enough from Kuhn's account . . . Such judgements are extensions of custom, which rely upon and affirm the body of accepted doctrine; an autonomous, unconditioned "reason," whatever that might be, has no role to play in judgements of this kind. Hence, if there is a role for "reason," it must be at times of paradigm change, when custom to an extent breaks down and the form and relevance of a sociological account of evaluation is less obvious. It is in revolutionary periods, when a choice between alternative modes of conventional activity is possible, that judgements sufficiently determined by logic and experience should appear, if at all. Kuhn, however, shows that they do not appear at such times, and that they cannot. (1982, 64)

Barnes appropriately divides his assessment of Kuhn into two parts, even though he claims that reason plays no role in either case. Decisions in normal science, that is *within a single paradigm,* are based on social custom, not on reason and evidence. Decisions made *between different or rival paradigms* are similarly social, according to Barnes, since

> logic and experience alone no more suffice than they do in normal science. There is no appropriate scale available with which to weigh the merits of alternative paradigms: they are incommensurable. To favour one paradigm rather than another is in the last

> analysis to express a preference for one form of life rather than another–a preference which cannot be rationalized by any non-circular argument. (1982, 65)

Would Kuhn himself be happy with a sociological reading of paradigm change–whether it came approvingly from Barnes or disapprovingly from Lakatos? Certainly not. He took up the matter in "Objectivity, Value Judgment, and Theory Choice" (1977b), in which he proposed five criteria for rational theory acceptance: *accuracy, consistency, scope, simplicity,* and *fruitfulness.* These are intended to be objective criteria for preferring one paradigm over its rivals. They are criteria that transcend any particular paradigm. We rightly prefer the theory or paradigm that makes more accurate predictions; we rightly insist on consistency, both internal consistency and consistency with other accepted theories; we rightly prefer the theory with greater scope, the one that accounts for more phenomena; and so on. These, says Kuhn, are the objective grounds for choosing among rival paradigms, grounds that are independent of the choices made.

The basis of objective choice is also the basis of explanation. Kuhn did not use the terminology of *reasons* and *causes,* but if he had, he would have said that the *belief* that paradigm P' is more accurate or has greater scope than paradigm P *caused* the members of the scientific community to abandon P and to accept P'. The change of belief is caused by these rational factors and is explained by citing them.

We may wonder whether Kuhn's criteria are adequate. They are certainly problematic–judgments of simplicity, accuracy, and so on are notoriously slippery. Kuhn admits as much:

> [I]f their specification is left vague, then such values as accuracy, scope, and fruitfulness are permanent attributes of science. But little knowledge of history is required to suggest that both application of those values and, more obviously, the relative weights attached to them have varied markedly with time and also with the field of application. (1977b, 335)

These are important concessions. Nevertheless, it is clear that Kuhn saw his own work in quite a different light than Barnes, the social constructivist, sees it, for Kuhn clearly thinks that paradigm change is a rational affair.

So much for the case of paradigm change. What about the process of normal science, that is, the process of belief change *within* a paradigm? Barnes and others see the process of normal science as similar to playing a game. The rules of any game, of course, are arbitrary. The permissible moves in chess, for instance, are conventional, just as those of checkers are. It is the accepted social conventions that make a particular move correct in one game but illegitimate in the other. There is no objective standard to which we can appeal beyond social custom. The rules of normal science, according to Barnes, work in the same way. There is nothing truly correct about the procedures of, say, classical mechanics, or of quantum mechanics; the rules of each are simply adopted and followed by the same sorts of social custom that the rules of chess or checkers are adopted.

So far the analogy is intelligible–if not plausible–but there is one crucial point that has been overlooked. A decision to play chess rather than checkers is itself conventional, perhaps a matter of taste. The choice of paradigms, however, is not–at least, on Kuhn's behalf, that is what we argued above. The details of normal science are forced upon us when we adopt a particular paradigm. When we consider *in isolation* a particular technique for measurement, or for calculating, or for model building, we may find that there is no reason that could be put forward for adopting it. But there is a good reason for choosing some particular paradigm, and once that paradigm is chosen, specific techniques of measuring and calculating come as part of the package. In this derivative fashion there is good reason for adopting those particular techniques, after all.

This is what is meant by Kuhnian holism. Consequently, *contra* Barnes, rationality within a paradigm (that is rational normal science), is indeed possible. It simply comes from choosing the best paradigm, then following the rules of that particular paradigm in doing normal science. So, if (following Kuhn) we take paradigm choice to be a rational activity, then normal science is a rational activity as well.

There are two stunning avowals in Kuhn's *Structure of Scientific Revolutions.* The first is the apparent abandonment of reason, though I suspect it to be inadvertent. At every turn after writing that book, Kuhn reaffirms the role of reason: decision making within a paradigm or between paradigms is indeed guided by good reasons. Unfortunately, he may not always successfully convince his audience of reason's role–just as Barnes hopes and Lakatos fears. It is clear, however, that Kuhn's intention is contrary to the reading of Barnes and others who use his work to underwrite social constructivism.

The second stunning thesis has never been recanted or even seriously qualified by Kuhn. This is his claim that science is not progressing toward any goal, such as the truth. Rather, theories are evolving–like Darwinian species–*from* earlier forms, but not *toward* any particular end. In this weak sense, and in that alone, theories (or species) can be better than their predecessors. Theories, according to Kuhn, do not correspond to reality even though they are getting progressively better. Interesting though that is, Kuhn's anti-realism is not our problem here. We're interested in the epistemology of science; we want to know about objective theory-choice, about the status of reasons and evidence, and about their effectiveness in determining belief. Of course, making Kuhn out to be a rationalist after all does not undermine sociological approaches to science. It merely robs them of a very prominent source of inspiration and support.

Bloor's Causal Principle

There is, as I've repeatedly stressed, a very wide spectrum of views within the social studies of science camp. Latour proposes an "actor-network" account; Pickering stresses the "mangle of practice"; classic SSK (sociology of scientific knowledge), as practiced by Bloor, Barnes, Collins, Shapin, Kusch, and others focuses on the social side of science and emphasizes interests and other sociological categories in explaining belief.

The emphasis on belief and belief change brings SSK close to traditional philosophy of science in this one key respect. Pickering and Latour, by contrast, are not much interested in belief, but instead focus on other things, such as practice. (Some philosophers of science do this, too; Ian Hacking, for example, emphasizes "intervening" over "representing.") There is a great gulf between social constructivists and most rationalist philosophers of science, but in the case of the practitioners of SSK, this gulf may be smaller than it seems. The similarities and differences to some extent turn on the details of their respective versions of naturalism.

Recall that naturalism says that the theory of knowledge should be modeled on the sciences. In particular, questions such as "What *should* we believe?, How *should* we evaluate our theories?" give way to "What *do* we believe?, How *do* we evaluate our theories?" In short, *normative* epistemology is abandoned in favor of *descriptive* matters of fact about sociology, psychology, and biology.

A sizable number of philosophers are naturalists.[1] Often they take their point of departure from Darwin. We are the products of natural selection, so it is very likely that some of our cognitive capacities result from adaptation; we are hard-wired to think in specific ways because these ways have survival value–not because we have evidence that these ways of thinking are correct.

Sociologically oriented constructivists would seem to differ from the philosophical naturalists only in their main source of explanatory principles: they draw on the concepts and categories of sociology (such as *interests*) rather than on the notions of evolutionary biology (such as *adaptation*). Otherwise, their outlook is remarkably similar–more similar, I suspect, than most philosophical naturalists would care to think.

Anti-naturalists needn't oppose all that naturalists proclaim. After all, some of our beliefs may indeed be the product of biological adaptiveness or the direct result of prevailing social factors. The principal point is only this: anti-naturalists (and I include myself in this camp) claim that there exist such things as reasons in the strongly normative sense. When these reasons are available to us, we ought to adopt the relevant belief, and when, in the light of these reasons, we do adopt the appropriate belief, then we do so because of those reasons. The anti-naturalist explains belief by citing the strongly normative reason for that belief–*reasons are causes.*

As we have already seen, the issue of naturalism comes up again and again. Now we can examine some of the main arguments proposed by champions of SSK for the nonexistence of reasons in this strongly normative sense.

First to something postponed from the last chapter, one of the most influential and naturalistically inspired arguments for the elimination of reason in the social studies of science. The first tenet of David Bloor's very influential strong programme is the *causal principle;* it implicitly rules out reason explanations altogether.

> [T]he sociology of scientific knowledge . . . should be causal, that is concerned with the conditions which bring about belief or states of knowledge. Naturally there will be other causes apart from social ones which will cooperate in bringing about belief. (Bloor 1976/91, 7)

This statement packs in much more than meets the eye. According to Bloor, the sociological approach to understanding science is part of a "scientific" approach to science; and we ought to be scientific. This is his expression of naturalism. (Let's ignore the harmless paradox involved in saying we *ought* to be naturalists.) Second, reason explanations are taken to be noncausal, hence, not natural or scientific. Bloor does not say this in so many words, but it is clearly implicit in all that he does. Finally, when he mentions "other causes" he presumably has biological and psychological causes in mind. Many of our beliefs, he allows, are not socially caused, but are due perhaps to our neural structures, and these have undoubtedly arisen through natural selection. (Sociobiology and evolutionary psychology attempt to explain our beliefs concerning incest, for example, on the basis of Darwinian adaptation, not on the basis of our having some sort of evidence that incest is objectively immoral, nor on the basis of it being merely a social taboo. The taboo arises because we are biologically determined to make it so.) Though Bloor's *Knowledge and Social Imagery* focuses on the social, Bloor takes a much wider view of what causes our beliefs–not so wide, however, as to include reason and evidence, at least not in the strongly normative sense as traditionally conceived.

I continue to use the term "rationalist" to signify someone who believes that reasons and evidence can cause belief. (A strict empiricist can be a rationalist in this sense; it should not be confused with any doctrine about *a priori* knowledge.) A rationalist need not believe that every belief is caused by a reason, only that some are. In short, rationalists hold that rational explanations are legitimate, in principle. Yet even in accepting this, there is still much room to disagree over the details. How should we characterize evidence and reason? Bayesians have one account, Popper has another, Lakatos, a third view, and so on. But this is a dispute over details and does not detract from the main point that reasons cause beliefs.

The basic idea is the same for all rationalists. How do we explain, for instance, the rise of quantum mechanics in the mid-1920s? We have already examined Paul Forman's (1970) sociological explanation: After the Great War, German scientists lost much of their prestige; Spenglerism was everywhere; the spirit of the times was anti-mechanistic. The scientists of the Weimar Republic created noncausal quantum mechanics to appeal to the general public and thereby regain their previous high social standing. By contrast, a rational explanation might claim that the old quantum theory was not a coherent set of principles; the new theory of Heisenberg and others accounted for a wide range of phenomena including the anomalous Zeeman effect, which had been the subject of much perplexity; consequently, scientists who worked in this field were won over by the successes of the new mechanics and completely accepted it.

It's important to note that there is no room for compromise here–at least, not for the anti-rationalist. Certainly rationalists can allow that *some* beliefs have been caused by social factors; but Bloor cannot allow that a single belief has been caused by reason and evidence (in the strongly normative sense). He cannot allow this any more than atheists can allow that perhaps there have been one or two miracles. For the serious naturalist like Bloor, there simply aren't such things as reason and evidence (as normally conceived), since they are not part of the natural world. In his own words, Bloor wants "to stop the intrusion of a non-naturalistic notion of reason into the causal story" (1976/91, 177).

There are a number of possible motivations for Bloor's position. For one thing, he insists that everything, including beliefs, be given a causal explanation and it seems obvious to him that reasons simply aren't causes; they are not the sort of thing that could have any effects. So, he insists, even if reasons and evidence did exist, they could explain nothing.

Are reasons and causes really such distinct kinds of things? Intuitions divide on this. Bloor was educated at Cambridge University at a time when many of the so-called ordinary language philosophers he might be reading explicitly held the doctrine that reasons are not causes.[2] But these same philosophers also held that reason-explanations of belief are quite legitimate. Bloor seems to have accepted half of this. Yes, reasons are a different kind of thing than causes, but no, reasons are not explanations–only causes can explain.

Perhaps it seems a plausible position to some, but it must be said that most philosophers today hold the contrary opinion: *reasons are causes* and *reason explanations are causal explanations.*[3] I should stress that this is not some newfangled view. In the history of philosophy, reasons for the most part have been taken to be a species of cause.[4] The causal principle (an event must be explained by citing its cause) does not have the force against rationalists that Bloor thought, since they, too, explain beliefs by citing causes. (More on reasons as causes below.)

Another motivation for skepticism about rationalist approaches to science comes from noticing how often "reason and evidence" are actually mere rationalizations. Too often pseudo-scientific theories are propped up by bogus evidence. Economists, psychologists, sociobiologists, and others offer a great deal of "evidential support" for their various accounts of how things are, but these theories may be little more than props for existing class, race, and gender relations.

A justified skepticism about a number of particular cases can, unfortunately, get easily generalized into skepticism about reason and evidence everywhere. Some of this skepticism is perfectly justified, but the leap to universal skepticism is not. To see this, consider once more Forman's explanation of the acceptance of quantum mechanics. Forman proposes that the Weimar scientists did not have evidence for their beliefs in quantum mechanics. He does think (implicitly) that they have evidence that adopting quantum mechanics will promote their (nonscientific) goal, which is to recapture their lost prestige. In short, reason and evidence are the causes of their nonscientific beliefs, according to Forman, but not of their scientific beliefs.

Many influential case studies have this form. (I mentioned this in the last chapter in connection with Bloor's symmetry principle.) On the surface they deny that evidence is at work, but the real structure of the argument is different. Usually the deep argument is that the participating scientists actually have nonscientific goals and that they reason in a perfectly ordinary manner to achieve these aims. Farley and Geison (1974) argue precisely this way. Pasteur, according to them, realized that he could better promote his reactionary politics by opposing spontaneous generation; consequently, he did so. Pasteur was not interested in the scientific goal concerning the truth of spontaneous generation, but he quite effectively used reason and evidence in pursuit of his political aims.

We can use these examples to make a distinction between *pragmatic reasons* and *evidential reasons,* as I will call them. A scientist has an *evidential reason* for believing, say, quantum mechanics, when she has reason and evidence for thinking quantum mechanics best achieves the scientific goals of truth, empirical adequacy, and so on. (We needn't be too specific about the details here, since these goals are themselves controversial.) A scientist has *pragmatic reasons* for believing quantum mechanics when she has reason and evidence for believing that by accepting the quantum theory, her nonscientific aims are best promoted. Put this way, we can then see that typical sociological accounts of science have implicitly explained a scientist's beliefs by using pragmatic reasons and have denied the existence or efficacy of evidential reasons.

A serious problem now arises. For how would we know that a pragmatic reason *P* serves the (nonscientific) goal *G*? Only because we have an evidential reason *E* that justifies the belief that the pragmatic reason *P* serves the (nonscientific) goal *G.* If there is an objection in principle to evidential reasons causing belief, then social constructivists who employ pragmatic reasons are as guilty as philosophers and traditional intellectual historians who cheerfully use the evidential sort. The only thing that distinguishes the evidential from the pragmatic is the nature of the goal they serve; and evidential reasons are needed even to see that pragmatic reasons might play a role in any particular situation. In short, one way or the other, we must *always* appeal to evidential reason.

One might think that pointing out this distinction would lead to the collapse of the sociological approach; that is, it could no longer be denied that reasons (in the strongly normative sense) are the cause of scientific beliefs. But no. Collins and Yearley (in a passage I discussed in the preceding chapter) would seem to be fully aware but still cheerfully ignore the point when they adopt their peculiar methodological stance:

> Natural scientists, working at the bench, should be naive realists—that is what will get the work done. Sociologists, historians, scientists away from the bench, and the rest of the general public

> should be social realists. Social realists must experience the social world in a naive way, as the day to day foundation of reality (as natural scientists naively experience the natural world). That is the way to understand the relationship between science and the rest of our cultural activities. (1992, 308)

In other words, scientists (*qua* scientists) should act as if evidential reasons are at work (even if they aren't). But the rest of us, standing back, should acknowledge the efficacy only of what I've been calling pragmatic reasons, since they are the ones that use sociological concepts such as interests to explain and understand scientific activity. This reflects a remarkable, if perhaps inevitable point of view, given the general principles of SSK. (Note, however, that it does violate the principle of reflexivity.) Collins and Yearley explicitly endorse it, and others implicitly embrace it in their treatment of various episodes in the history of science.

The first problem with this methodological principle (as I mentioned in the previous chapter) is that it makes for a kind of schizophrenia. Scientists, while they are being scientists, are allowed–indeed, they are encouraged–to think there are good evidential reasons for the beliefs they adopt. At the end of the day, however, when they take their lab coats off and reflect on what they've done, they disown all. None of their beliefs is the product of rational considerations; everything is rather the product of social factors.

There is a second reason for doubting the plausibility of the Collins-Yearley principle. It assumes a sharp distinction between nature and society. According to them, there would be a great difference between "The proton has mass *m*" (which is a social construction, not an objective fact) and "Pasteur wants fame and fortune" (which is not a social construction, but rather an objective fact about social reality). What about "Pasteur wants lunch"? There's certainly a social side to this, since "lunch" involves all sorts of social conventions. How about "Pasteur wants to eat"? This is getting further away from the social. Finally, "Pasteur has low blood-sugar" seems to be on the nature side of the divide–but it's hard to say. I suspect a continuum of cases is possible here, and the line between social realism and constructivism may be hopelessly blurred.

Aside from the terrible schizophrenia this induces and the arbitrariness of the natural-social boundary, there is a third, much bigger problem. This principle makes a mockery of a certain class of case studies done in the SSK mold. In the explanations of Forman and of Farley and Geison, the working scientists are themselves *not* realists, but rather are at some conscious level aware of their own interests and the relevant social factors. The scientists consciously adopt their beliefs because of the social situation–on the basis of pragmatic (interest-serving) rather than evidential (truth-serving) reasons. That the scientists in question were fully aware of their social situation and were trying to improve it, as opposed to being realists who were trying to figure out how nature works, is crucial to the sociological explanation of their behavior. For years, these case studies have been cited as brilliant supporting illustrations of SSK. This claim now rings hollow.

Reasons Are Causes

Why would we doubt that reasons are causes? To start with, there are strong intuitions that say reasons can't be. People often produce after-the-fact rationalizations and try to pass them off as the evidence for their beliefs. Obviously, a reason, however good in its own right, that came to light only *after* a belief was adopted is not a cause of that belief. Often with apparently very good reasons staring us in the face, we still don't adopt the appropriate belief–a loved one is dying of cancer, but believing it is too painful. Sometimes we adopt the correct belief when the evidence is available, but not because of the evidence. I adopted my beliefs about the nature of a vacuum not because of the detailed, correct explanation of my elementary-school teacher, but because I had a crush on her. Self-deception, weakness of will, and other such considerations can easily lead one into skepticism with regard to the causal efficacy of reason.

There are more sophisticated considerations for doubting reasons are causes, stemming from naturalism. As I mentioned above, naturalism is the view that everything can be accounted for by means of natural science. (Physicalism is an extreme version of this.) The crucial thing that naturalism rules out is norms. Values and norms are eliminated, explained away, or reduced to natural entities and processes. Bloor's "science of science" is an expression of this sort of naturalism. The crucial thing about reason and evidence (as normally conceived) is their normative aspect. To say that *R* is evidence for *P* or a reason to believe *P* is to say that one *ought* to believe *P* (unless, of course, there is even stronger evidence to the contrary). The easy way to eliminate the "ought" is to eliminate reason and evidence outright, and to explain beliefs with other kinds of causes.

However, the foregoing considerations all seem outweighed by the simple structure of a reason-explanation, which makes it apparent that reasons are indeed causes. We want to know why, for example, *A* believes *P*. Explanations typically have this form:

1. Evidence *E* implies the truth of *P* (relative to *A*'s background assumptions).
2. *A* was aware of evidence *E* and aware that *E* implies *P*.
3. *A* adopts the known consequences of her beliefs (other things being equal).

∴ *A* adopted the belief *P*.

This is clearly a causal explanation. And the reason (i.e., the evidence *E*) is the cause of the belief that *P*. Strictly, it is not the reason *E* that causes the belief *P*; it is *A*'s belief that *E* holds that does the causing. This is captured by saying *A* is "aware" of the evidence. It is a bit tedious unpacking all of this, but a fuller analysis might go something like this: There is a state of affairs in the world correctly described by proposition *E*. *A* perceives this state of affairs–a causal process–which results in *A* believing the proposition *E*. The proposition *E* implies the proposition *P* and *A* is aware of this logical relation. *A* has the disposition to accept elementary, known consequences of prior beliefs. Thus, *A* accepts the proposition *P*. We abbreviate this by simply saying that *the reason caused the belief.*

Reasons Sociologized

In talking about reasons and evidence I have repeatedly added the caveats "as reasons are normally understood" and "in the strongly normative sense." This is because many champions of SSK would deny that they are hostile to reason and evidence. They would say, on the contrary: "Of course reason and evidence are at work, it is just that we have to take the *context* of such factors into account, and that context is *social;* there is no such thing as reason and evidence, *simpliciter.*"

> What counts as an "evidencing reason" for a belief in one context will be seen as evidence for quite a different conclusion in another context. For example, was the fact that living matter appeared in Pouchet's laboratory preparations evidence for the spontaneous generation of life, or evidence of the incompetence of the experimenter, as Pasteur maintained? As historians of science have shown, different scientists drew different conclusions and took the evidence to point in different directions. This was possible because something is only evidence for something else when set in the context of assumptions which give it meaning–assumptions, for instance, about what is *a priori* probable or improbable. If, on religious and political grounds, there is a desire to maintain a sharp and symbolically useful distinction between matter and life, then Pouchet must have blundered rather than have made a fascinating discovery. These were indeed the factors that conditioned the reception of his work in the conservative France of the Second Empire. "Evidencing reasons," then, are a prime target for sociological inquiry and explanation. There is no question of the sociology of knowledge being confined to causes *rather than* "evidencing reasons." Its concern is precisely with causes as "evidencing reasons." (Bloor and Barnes 1982, 28f)

At first glance this passage makes it seem that reasons are playing an acknowledged role. Yet this illusion quickly fades when we make a distinction between *reasons as causes* and *"reasons" as rhetoric.* In either case Bloor and Barnes are quite right to point out the importance of context or background, but they do not accept a causal role for reason; rather, they assert a causal role for reason-talk.

First, the point about context, which is certainly right: reasons are only reasons in a context. The most trivial examples will make this clear. Suppose Mary believes *Q.* Her reason for believing this is *P* (i.e., she believes *P*). Clearly something is missing, since *P* is not by itself a reason (not a cause) for believing *Q.* Let us further suppose that Mary believes that $P \rightarrow Q$ and that she readily sees elementary logical consequences. By filling in the context–Mary's background beliefs that hitherto went unstated–we can fully account for her belief. In detail our explanation runs: Mary believed P and she believed $P \rightarrow Q$, and she further realized that Q is a logical consequence of her beliefs. Thus, she adopted the belief Q.

In a different context, say, in which her background belief was $P \rightarrow \sim Q$, then she would not have adopted Q, but rather its negation. Thus, in one context P is a reason for Q, but in another P is a reason for $\sim Q$. There is nothing paradoxical about this. Nor is there anything particularly social about it.

Second, let's look at *reasons* vs. *reasons-talk.* Barnes and Bloor say they are concerned with "causes as 'evidencing reasons.'" But what do they mean, and what has this to do with contexts? The examples they give of "contexts" are all political. These include background beliefs and assumptions, to be sure, but they smack of social goals and interests rather than ordinary scientific beliefs. Thus, the context in which Pouchet is considered is the politically and religiously conservative France of the mid-nineteenth century. This, however, is the social context. The epistemic context–the context that matters when it comes to reasons–is ignored. This latter context would include background assumptions from chemistry, biology, and so on. It is the "scientific" assumptions that play the relevant causal role in a so-called rational explanation of belief and belief change. Those assumptions from chemistry, biology, and so on may themselves be highly conditioned by social factors, but that is not the issue. It is the *scientific,* not the social context, that one must appeal to in using contextualized reason-explanations. This is what Barnes and Bloor fail to do.

When Barnes and Bloor use "evidencing reasons" in an explanation, they are claiming (correctly in some cases, no doubt) that some scientists use a particular proposition (e.g., "Pouchet found living matter in the container") one way (to show spontaneous generation), whereas others use it in a different fashion (to show Pouchet was incompetent). But in each case the scientist's use is rhetorical. If Barnes and Bloor are right, then reasons are not causes; rather, it is the rhetoric of reason and evidence put at the service of nonscientific interests that is doing the real causing.

The example I just used might give rise to a worry. The belief in P and in $P \rightarrow Q$ were given as reasons for the belief in Q; but one might well wonder where P and $P \rightarrow Q$ come from. Perhaps they are themselves full of social content? It might be possible to find reasons for belief in P, but clearly a regress is shaping up. Moreover–and we might as well admit it outright–no belief is directly evident, either. Does this mean that none of our beliefs is justified? Perhaps it does, but I won't pursue the intricacies of general skepticism here. A simple distinction will suffice.

We must distinguish between *belief* and *change of belief*. The regress problem gets in the way of our saying that Mary's belief in P or in $P \rightarrow Q$ is rational. We can, however, say her change of belief from, say, $\sim Q$ to Q is indeed a rational change of belief. Given her background assumptions, she made the correct move in rejecting her old belief $\sim Q$ and adopting Q in its place. When we try to explain beliefs in terms of their rational merits, it is actually change of belief that we are trying to explain. Change of belief is for the most part what we're interested in within the history of ideas. When we wonder why people believed T, we are often asking the more complicated question, "Why did they abandon T' and adopt T in its place?" This simple point obviates the regress problem.

Is There a Middle Ground?

No. But let me first explain what the middle ground is. I've been arguing that reason is part of the causal story; evidence, rational factors, and good reasons can be the causes of beliefs, and they explain why those beliefs came about and why they are sustained. Bloor and other SSK champions completely deny this, at least when reason is understood in a strongly normative way. (When reason is sociologized it loses its normative force.) It is crucial to Bloor "to stop the intrusion of a non-natural notion of reason into the causal story" (Bloor 1976/91, 177). This is because any nonnaturalized notion of reason is illegitimate. Genuine norms with the causal power to create and sustain belief are for Bloor on a par with unicorns–utterly mythical. They are not part of the natural world at all; they simply don't exist.

There is no room for compromise here. Either the world has genuine norms that play a causal role or it does not. But some have tried to have it both ways. Michael Friedman, for example, is highly critical of SSK and especially of its use of philosophers such as Wittgenstein to support their various relativistic theories of knowledge. When it comes to the role played by norms, he wants to adopt a middle view. In response to Bloor's banishing of all nonnaturalistic notions of reason, Friedman writes:

> But this line of thought rests on a misunderstanding. All that is necessary to stop such an "intrusion" of reason is mere abstinence from normative or prescriptive considerations. We can simply describe the wealth of beliefs, arguments, deliberations, and negotiations that are actually at work in scientific practice, as Bloor says above, "without regard to whether the beliefs are true or the inferences rational." In this way, we can seek to explain why scientific beliefs are in fact accepted without considering whether they are, at the same time, rationally or justifiably accepted. And, in such a descriptive, purely naturalistic enterprise, there is precisely enough room for sociological explanations of why certain scientific beliefs are in fact accepted as the empirical material permits. Whether or not philosophers succeed in fashioning a normative or prescriptive lens through which to view these very same beliefs, arguments, deliberations, and so on, is entirely irrelevant. In this sense, *there is simply no possibility of conflict or competition between "non-natural," philosophical investigations of reason, on the one hand, and descriptive, empirical sociology of scientific knowledge, on the other.* (Friedman 1998, 245, my emphasis)

Friedman seems entirely on Bloor's side when it comes to explaining the actual events of history–it is irrelevant whether Newton or Maxwell or Darwin actually had *good reason* to believe what they believed. What *caused* them to adopt their particular views is the type of *natural, nonnormative* factor that Bloor and other SSKers cheerfully embrace. They see eye to eye on what has actually happened and the causes of it. Friedman only differs from Bloor in saying that we can *also* correctly describe certain beliefs as rational. Good reasons also exist; it's just that they play no causal role in actual history.

I see two significant problems with Friedman's account. If there are going to be such things as evidence and good reasons (in the strongly normative sense), then it is something of a mystery why they should play no causal role in the history of belief. If they are totally disconnected from everything else, then Bloor is surely correct to dismiss them entirely. Otherwise, it's a bit like saying there really are unicorns, but that unicorns are completely undetectable entities that do not causally interact in any way at all with other objects. It's logically possible that there are such unicorns, but good sense and Occam's razor tell us to disbelieve their very existence.

A second problem is related to the first. If pressed, Friedman and other champions of this view might resort to pointing out some embarrassing facts. They will note how bizarre Newton, for example, really was. He held all sorts of kooky beliefs on all sorts of subjects, especially in religion and alchemy. It's hard to believe that he suddenly became a paragon of rationality when doing mechanics. Real life is messy and complex; all sorts of factors would have been at play in Newton's thinking. It's irrelevant whether or not he was "rational," according to this line of thought. What matters to us, says Friedman, is that we can *reconstruct* (rationally) the historical episode and see in retrospect that Newton's theory (i.e., the laws of mechanics and the gravitation law) was warranted in those circumstances.

But is *our* belief that Newton's theory was warranted itself warranted? Is our belief just another event brought about by various natural causes, as Bloor would maintain? Or is our belief itself a rational belief, caused by the evidence that is available to us? If the former, then Friedman's thought that we (current historians) can get a grip on norms (even though they play no role in the causal story of the past) is simply wrong. If it is the latter, then we are capable of having our beliefs be caused by evidence. And that means that Newton is capable, too. Thus, reason can indeed be part of the causal story, contrary to Bloor and Friedman alike. There is more to understanding the history of scientific beliefs than merely recording (however systematically) the totality of natural causes. There are nonnatural causes of belief, too, namely, reason and evidence.

At one point Friedman remarks: "It is not that the philosophical tradition sets up a competing model for causally explaining the actual historical evolution of science" (1988, 250f). On the contrary, that is precisely what I'm trying to do.

Underdetermination

One source of the rejection of reason in the sociology of science stems from naturalism and the conviction that reason (in the strongly normative sense) can't be part of a naturalistic outlook. The other main source stems from a cluster of philosophical problems such as "underdetermination," which has proven to be a very influential consideration. Sociologists of knowledge–like the rest of us–are opportunists. They would probably say their main case rests on the success of their detailed case studies of various episodes in the history of science, but when philosophy inadvertently offers a helping hand in the form of the underdetermination thesis–grab it! So, let's now turn our attention to it. In a nutshell, the argument from underdetermination says that even if there are such things as reasons and evidence, they are *not sufficient* to cause belief. First, what is underdetermination?

Almost two thousand years ago Ptolemy proposed a system of the world. The sun, moon, stars, and planets all move in circles around a stationary earth. In many cases the circular motion is actually motion along circles on circles (called epicycles). An appropriate arrangement of epicycles could yield almost any possible motion, so that viewed from the earth the appearance of, say, Mars would be empirically correct. That is, Mars would be at location x at time t just as Ptolemy's theory says it would.

Copernicus, 1,400 years later, offered quite a different account of things. He put the earth and other planets into orbital motion around a stationary sun. The difference in the appearances of Mars stem from Mars and ourselves moving along our respective orbits. Copernicus, too, could accurately predict the appearance of Mars. The problem of deciding which theory is correct stems from the fact that both accurately account for the apparent motion of the individual planets, both accurately predict eclipses, and so on. There was no hope of getting outside the solar system to have a "God's-eye-view" to determine which theory was really true. The two theories seemed to be empirically equivalent and empirically correct. There was no empirical evidence available at the time of Copernicus that would determine which (if either) was true. In other words, they were *underdetermined* by the evidence.

This is not a perfect example of underdetermination. By the early nineteenth century an empirical difference became available in the form of stellar parallax, resulting in the slightly different appearance of the stars in, say, January and July, because the earth is on different sides of its orbit. Nevertheless, the example gives the idea of what the phenomenon of underdetermination is and of why it could lead to philosophical problems. Because there are no known perfect examples, we have to discuss the problem rather abstractly.

There are many variations on the problem of underdetermination, but a fairly simple version will suffice here. Consider a theory T that does justice to the available evidence, which we'll take to be a body of observation sentences, $\{O_i\}$. (This will include statements such as "Mars is at location x at time t.") And we'll take "doing justice" simply to mean explaining the data by entailing it, i.e., $T \rightarrow \{O_i\}$. (If the theory is true then the observation statements must be true as well.) Thus, the problem of underdetermination arises from the fact that there will always be another theory (indeed, infinitely many others) which will do equal justice to that very same data; that is, there is always another theory, T', such that: $T' \rightarrow \{O_i\}$. So how do we choose? The available evidence does not help us to decide; theories are underdetermined by the data.

I said a simple version of the problem would suffice. Among the simplifying assumptions just made are these: We have assumed that the observation sentences alone constitute the evidence; we have taken entailment to be how a theory accounts for and does justice to the evidence; we have assumed the evidence is expressible in a neutral way, that is, that the same observation sentences, $\{O_i\}$, are entailed by the various rival theories. All of these assumptions could–and should–be criticized. Kuhn, more than anyone, has taught us that. But let's overlook this for now since the nub of the underdetermination problem exists in this simplified form, as well as in more subtle versions.

Typically, traditional intellectual historians explain beliefs by citing the evidence that supports them. Why did Newton believe the theory of universal gravitation? Because of the evidence he had for it. Why did Darwin believe in evolution? Because of the evidence he had for it. The approach is not only standard for historians of ideas, it is entirely common-sensical. Why do we believe that it will rain tomorrow?, or that all the pizza is gone? Because of the *evidence* we have for these beliefs.

Yet if there are alternative theories that do equal justice to the available evidence, why are our current beliefs held and not one of the alternatives? We cannot appeal to the evidence, for there is no evidence that is sufficient to single out one among the indefinitely many alternative theories, a theory that is uniquely able to account for that evidence. So there must be something else that explains why the belief has been adopted. Obvious candidates are social or psychological factors that cause the adoption of one out of the many possibilities. Thus, according to Forman (1971), it was not the available evidence that made the Weimar scientists adopt nondeterministic quantum mechanics in the mid-1920s, but rather their interest in regaining their social position and prestige, lost in the disastrous Great War. According to Shapin in his famous study of class war and phrenology in early nineteenth-century Scotland (Shapin 1975), it was not the available evidence that made the middle class of Edinburgh adopt phrenology, but rather their desire to attack the upper classes and acquire a larger share of political power. Similarly, it would be argued, Newton and Darwin and every other scientist have held their beliefs because of social factors, since the evidence available to them was never enough to justify choosing their theories over all the alternatives.

Naturalists such as David Bloor are deeply skeptical of "evidence" at the best of times; and insofar as it does seem to play a role, "evidence" can, he insists, be given a sociological interpretation. (This we saw above.) There is, he thinks, nothing objective answering to it. The underdetermination argument, however, says: Even if there is such a thing as evidence, it is not sufficient to explain belief; we must appeal to social (or other natural) factors to explain why one particular theory was adopted rather than some other that is evidentially equivalent. So, not only are rationalists (i.e., those who try to explain beliefs by appeal to reasons) working with a spurious entity, evidence, but even if evidence did exist, it cannot do what it is touted to. The argument seems to have some bite.

Let $\{O_i\}$ be our body of data, the evidence, and let $T, T', T'', \ldots$ all equally account for the data. We shall further suppose that T is accepted, and we want to know why. The sociologist of knowledge posits an interest, *I*, which is the cause of *T* being adopted; i.e., *T* serves interest *I* and that is why *T* was accepted. The interest explains the belief.

Thus, as well as accounting for $\{O_i\}$, a theory must also do justice to the interest *I*. However, underdetermination works here, too.[5] There are indefinitely many theories, *T*, T^*, T^{**}, . . . which will do equal justice to $\{O_i, I\}$. So, we can ask again: Why was *T* selected? We don't yet have an explanation.

Now the sociologist can take either of two routes: The first route is to propose a second interest, I_2, to explain why *T* was chosen over its rivals. But this leads to an obvious regress. Undoubtedly there are only a finite number of interests a person could possibly have; so eventually the explanatory power of interests will stop, leaving it unexplained why *T* was chosen over indefinitely many alternatives that could do equal justice to $\{O_i, I_j\}$ (where $1 \leq i < \infty$ and $1 \leq j \leq n$).

The second route is to deny that there are really indefinitely many alternatives. The sense in which indefinitely many alternative theories *exist* is a logician's Platonic sense, not a practical one. In reality, there are usually only a handful of rival theories from which to choose, and from these one will best serve the scientist's interests. It is a small irony that sociologists of science who stress the importance of context conveniently forget that in real life theories are accepted or rejected in contexts that include only a very few rival theories.

This second route would indeed be sufficient for an explanation, but if there are only a very small number of theories to choose from, the evidence $\{O_i\}$ will almost always suffice to decide among them. In other words, there is no practical problem of underdetermination, so there is no need to appeal to something else, an interest, to explain the adoption of the theory *T*. Underdetermination, consequently, cannot be used to justify a sociological approach to science. Reason and evidence can and do explain belief.

In the social studies of science reason and evidence play a diminished role. They are eliminated outright when understood in a strongly normative sense, and even when "naturalized" they still play only a negligible role. Naturalism and underdetermination are the two main sources of support for this. As we have seen, the case for slighting the causal effectiveness of reason has not been made. Indeed, quite the opposite is true: Reasons and evidence are indeed causes and thereby explanations of beliefs.

While still on the topic of underdetermination, I should mention, if only briefly, the attitude of some of the logical positivists. Various authors–in displays of unparalleled ignorance–describe positivists as conservatives–scientific and political. On the contrary, Otto Neurath, for example, not only accepted the underdetermination thesis, but cheerfully embraced it for political ends: whenever we have rival theories and evidence does not decide between them, we should choose the socially progressive theory.[6]

Before quitting, let's look briefly at reason in its most dramatic and effective role–in thought experiments.

Thought Experiments: A Challenge

There are social constructivist analyses of theories (why this one was accepted and that one rejected), of experimentation (why this practice was adopted and that one abandoned), and a number of other features of science. But there has never been a constructivist analysis of thought experiments, neither in general nor of any particular thought experiment. Why is this? Thought experiments certainly play a big role in science, so they should be a target for constructivist analyses. It might be that no constructivist has bothered to think about them, but if prodded into action they could give perfectly plausible accounts–at least as plausible as social constructivist accounts of other aspects of science. I suspect, however, that thought experiments will prove a greater challenge for constructivists than any other aspect of science. The reason is simple: here, more than anywhere else, reason is clearly at work. Before I explain this, let's look at my favorite example, Galileo's wonderful argument in the *Discorsi,* which shows that all bodies, regardless of their weight, fall at the same speed (*Discorsi,* 66f).

It begins by noting Aristotle's view that heavier bodies fall faster than light ones. (We denote this by $H > L$. The symbols H, L, and $H+L$ are used ambiguously to denote the objects of different weights and to denote their speeds when falling.) We are then asked to imagine that a heavy cannonball is attached to a light musketball. What would happen if they were released while joined together?

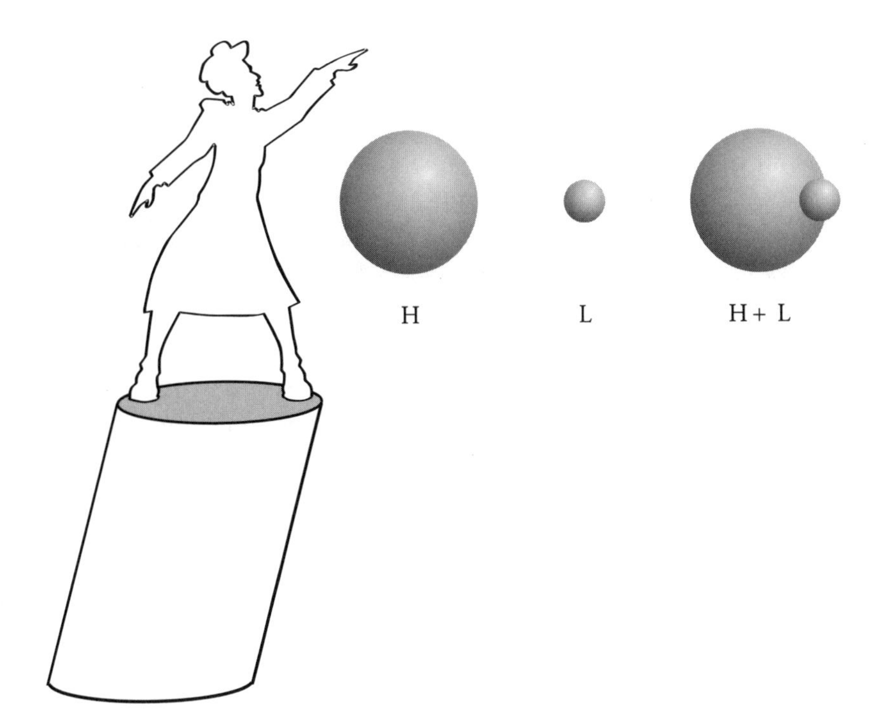

Figure 7 Galileo's thought experiment showing that all bodies fall at the same rate.

Reasoning in the Aristotelian manner leads to an absurd conclusion. First, the light ball will slow up the heavy one (acting as a kind of drag), so the speed of the combined system would be slower than the speed of the heavy ball falling alone ($H > H+L$). On the other hand, the combined system is heavier than the heavy ball alone, so it should fall faster ($H+L > H$). We now have a logical contradiction, namely, the absurd consequence that the heavy ball is both faster and slower than the even heavier combined system. Thus, the Aristotelian theory of falling bodies is destroyed. But the question remains, "Which falls faster?" The correct answer is now plain as day. The paradox is resolved by having them all fall at the same speed ($H = L = H+L$).

There is no consensus, even among those who reject social constructivism, on how thought experiments work. Some offer Platonist accounts, others are staunch empiricists, and there is a large spectrum in between.[7] But one thing they all have in common is that conceptual matters reign supreme. Thought experiments are carried out by manipulating concepts. Nothing else plays a significant part. The kinds of considerations to which social constructivists typically appeal in their accounts of theory-choice or experimental practice are wholly absent in the process of thought experimenting.

I'll end with an explicit challenge: *No account of thought experiments along social constructivist lines can be given.* Of course, I don't really expect all attempts to end in a failure that is manifestly obvious for all to see. I'd even be quite interested to see how an attempt might go. However, I do expect it to be very difficult, if not impossible, since what needs explaining is so purely conceptual, so intimately tied to reason.

8

The Democratization of Science

A central theme of the science wars has been *the democratization of science.* Is it a good thing? How could it be brought about? What does it even mean? There are several things that come to mind as to the meaning of the democratization of science: there is democratization of research funding, democratization of the choice of research problems, democratization in the form of a wider representation of social groups, and democratization of policy issues concerning which technologies to use or avoid. All of these matter greatly. The principal focus of this book, however, has been on cognitive questions—to what extent, if any, is science objective? In asking questions about the democratization of science, our chief concern is with objectivity. However, other aspects of democratization will inevitably arise along the way.

When it comes to ordinary politics, we *all* like democracy—or at least say so—but there's no consensus at all in science. One opinion, common among scientists and philosophers since Plato, says that science *is not* and *should not* be democratic. Truth is truth, evidence is evidence—it's not a matter of taking a vote. Comical examples could be cited: the democratically elected State of Illinois legislature (about a century ago) passed legislation to the effect that π is a rational number; and (less comically) some democratically elected school boards (again, in the United States) insist that "creation science" be taught alongside Darwin. Of course, critics who want science to be democratized don't have such silly examples in mind. But can they escape them?

There's much talk of a need for democratized science by left-wing critics, but little or no detail is offered. Let's briefly look at some of the sketchy proposals that have been put forward.

Democratic Science from a Democratic Society

Traditional Marxism holds that social problems stem from class conflict. There is a great deal of truth in this–yet it can't be the whole story. A number of social problems are greatly exacerbated by class conflict, but do not wholly arise from it. Would racism and sexism disappear if class conflict ended? Not likely. Racism and sexism seem to have a life of their own. How might science itself fare in a more democratic society? Would it change for the better?

In numerous writings Hilary Rose and Stephen Rose have made the case for a more democratic science along Marxist lines. Their claim seems to be that we will have a more democratic science *when and only when* we have a more democratic (i.e., classless) society.

> Certain types of society therefore do certain types of science; they ask particular questions of nature. Ancient Babylonian religion demanded the accurate prediction of heavenly events and Babylonian science was largely devoted to the intensive study of astronomy. The emergent capitalism of the industrial revolution in Britain required technological advances in power generation and physicists studied the laws of thermodynamics and conservation and transfer of energy . . .
>
> But the negative corollary of this relationship between science and society holds true too; that is, that in certain societies certain types of science are *not* done . . . [For example,] the withering of physics in Nazi Germany and of genetics during the Lysenko period in Russia.
>
> In this sense society gets the science it neither demands, nor deserves, nor needs . . .
>
> It is neither avoidable nor wrong that such constraints should exist. The questions we have to ask, in the long run, must take cognisance of them. They are: what sort of science do we want? how much of it do we want? who should do it? how should they and their activities be controlled? But the fundamental question underlying all these activities is: what sort of society do we want? (Rose and Rose 1970, 244f)

The Roses note that "This thesis, that science springs from the economic base of society, has been central to the Marxist analysis of the history and philosophy of science. In Britain this type of analysis began to be made in the 1930s by such scientists as Desmond Bernal, J. B. S. Haldane, and Joseph Needham" (ibid., 244). Bernal, Haldane, and Needham were among a wholly admirable group of top-notch scientists deeply concerned with social problems, especially those connected with science. Their main consideration was that science is constrained under capitalism, but under socialism they believed it would flourish, and flourish in a way that genuinely served society.

It's impossible not to sympathize with the view put forward by the Roses and earlier by Bernal and the other British Marxist scientists, but it's also impossible to find it wholly believable. Like racism and sexism, undemocratic aspects of science seem to have a life of their own. It will take more than a change in society toward social and economic equality to bring about the desired changes in science itself. Moreover–and this is surely the crucial point–we can't wait for a more democratic society. Changing science for the better now might even help to promote a more democratic society in the future.

Negotiating the Facts

Andrew Ross would like to see science democratized in the sense of taking much of the power out of the hands of "experts" and giving greater input to "local" people who have some relevant interest.

> [T]he remoteness of scientific knowledge from social and physical environments in which it will come to be measured and utilized is as irrational as anything we might imagine, and downright hazardous when it involves materials that can only be properly tested in the open environment. The unjustified conferral of expertise on the scientist's knowledge of, say, chemical materials, and not on the worker's or the farmer's experience with such materials is an abuse of power that will not be opposed or altered simply by demonstrating the socially constructed nature of the scientist's knowledge. That may help to demystify, but it must be joined by insistence on methodological reform–to involve the local experience of users in the research process from the outset and to ensure that the process is shaped less by a manufacturer's interests than by the needs of communities affected

> by the product. Such methodological reform will lead from cultural relativism to social rationality. (1996, 4)

In the Ross model democratization comes through sharing power; the local farmer should share decision-making input with the expert chemist. Although Ross doesn't tell us the details, we can easily imagine a couple of ways that things might go. One way of sharing power would have the farmer informing the chemist of facts of which the chemist had previously been unaware, facts that could come to light by being involved in agriculture in the way that only the farmer is. Not surprisingly, it is extremely easy for a traditional champion of science to completely sympathize with this. After all, there's nothing in the orthodox outlook that says nature is easy to comprehend; the more sources of evidence available, the better our understanding will eventually be.

A second way to interpret the Ross power-sharing model of democratic science is to see the chemist and the farmer "negotiating" what is to be done, as if they have competing interests and it's a matter of achieving a fair balance. This would clearly be a democratization in the constructivist mold. Neither scientist nor farmer is trying to get at the objective truth of the matter (since it doesn't exist, anyway), but each is rather trying to maximize his or her interests to the fullest. Democracy plays a role in that they negotiate on a level playing field. Ross, antirealist that he is, probably has something like this second version in mind.

One of the problems with this second version is that it sees only input coming from the expert chemist and the local farmer. It fails to note that there may be more legitimate interests involved. What about consumers? If the issue is pesticides and fertilizers, then those who eat the food the farmer grows have a very big stake in what happens. Ross pits the local farmer against the expert chemist, but the interests of both may be contrary to the interests of consumers. Farmers might worry about the pesticides they inhale, to which consumers (*qua* consumers) are indifferent. On the other hand, farmers (*qua* farmers) are indifferent to the pesticides ingested from their food, whereas consumers are rightly worried. It is even possible that the expert chemist acting alone might be a better protector of consumer health than the more democratic duo of chemist and farmer.

The second thing that Ross overlooks in his example is that there are relatively few chemists and even fewer manufacturers, but many farmers and a great many more consumers. Throughout his discussion he uses the singular "farmer" and singular "chemist," implicitly suggesting that these two could (at least in principle) sort things out. If ever there were a case for representative rather than direct democracy, here it is. (I'll elaborate on this point below.) The chemist can't possibly negotiate with each and every farmer, but input from a representative farmer is perfectly feasible.

PRIVILEGED SUBJECTS

One of the aims of feminist science, especially feminist social science, is the empowerment of women–an ideal of democracy. This has led to a great deal of interest in method. What do feminist social science and especially feminist method look like? One popular answer says that research subjects get to interpret their own experience. This may sound innocuous, but is actually far-reaching and quite controversial. The general idea has spread to anthropology as well, where the indigenous people's interpretation of their own experience is given not just serious, respectful consideration, but is privileged and primary. Self-image is taken to be much more than a datum.

Stanley and Wise (1983) provide one of the most important and influential versions of this methodological outlook. The perspective of the subject of research is to be taken at face value and they maintain an unqualified "insistence on the validity of women's experience" (1983, 135). What women say about themselves is not to be questioned, not to be interpreted by the researcher, not to used to build up any sort of general theory or framework.

This can lead to a disturbing situation. "If a housebound, depressed, battered mother of six with an errant spouse says she's *not* oppressed, there's little point in us telling her she's got it wrong because of the objective reality of her situation." On the contrary, her version is indeed the "*truth* for her" and we must refrain from "attempting to impose our reality" on her (Stanley and Wise 1983, 112f).

Such relativism leads to immediate problems. It's logically possible to treat men and women differently from a methodological point of view (i.e., take women's experience at face value but not men's), but such an asymmetry would seem rather arbitrary and very hard to justify. If treated the same, however, then the wife-beating, rapist's version of what happened ("It was obvious that she was asking for it.") would similarly have to be taken at face value and left unchallenged–which is absurd.

The Stanley and Wise model of social science research is explicitly based (in part) on consciousness-raising techniques. *Ms. Magazine* published a famous set of guidelines for consciousness raising in 1972. The primary rule is never to challenge or even advise another woman in her account of her own experience; she can only do that herself. There is one important difference. Consciousness raising has the explicit aim of changing people's outlook, and would be considered a failure if it didn't. Stanley and Wise seem to resist this, even though they have in other respects the usual feminist goals. There would seem to be more than a little tension in their outlook.

At first blush the repeated insistence that researchers take women's experiences seriously seems merely good and proper ethical conduct. Traditional researchers would likely agree with the injunction, but would take it to be part of research *ethics,* not part of research *epistemology.* For example, in most cases it makes no difference to the experimental results whether research animals are anaesthetized before their livers are examined, but it would be horrifically cruel to inflict needless pain. This, perhaps, is the crux of the difference, and one way of seeing the force of the privileged-subject methodology is to see it as rejecting the distinction between ethics and epistemology. After all, the aim of feminist social science is, among other things, to empower women. This is why a link between politics and knowledge is sought in the first place.

The idea of taking the subject seriously has become a major component in much recent anthropology, as well. The general idea is the same: it is not the anthropologist, but the indigenous people who get to interpret their own culture. To some extent, this way of thinking is revealed in a confrontation that followed the Sokal hoax.

The early aftermath of the Sokal affair included a debate at New York University on October 30, 1996 between Sokal and Andrew Ross, both New York University professors.[1] Sokal stressed that he is not out to defend science, but to attack an approach to social issues that is related to science he thinks utterly misguided. The best way to tackle social issues, Sokal claims, is by being a critical rationalist.

He raised an example of what he took to be sloppy thinking, citing an anthropologist who remarked that western cosmologies and those of Native Americans were equally valid. The first tells a story something like this: After the Big Bang came the formation of stars and planets, followed by the evolution of life on Earth. Eventually some members of *Homo sapiens* migrated out of Africa and across Asia, then about ten to forty thousand years ago they crossed the Bering Strait into North America. That's how Native Americans came to be here.

The other cosmology (and here I simplify even more than in the first case) says something like this: Native Americans emerged long ago from a subterranean spirit world. They did not migrate from some other part of the globe, but sprang up from the very ground beneath us. Since their beginning they have always been here.[2]

These accounts can't both be right–though both could be wrong. How, Sokal wonders, could any sensible person endorse them both? This, he thinks, exemplifies the sloppy thinking to which postmodernism leads, however well-intentioned.

The ensuing discussion was most revealing.[3] Members of the audience challenged Sokal, asking him what right he had even to pose this problem. Ross objected that Sokal was "putting Native Americans on trial" with his example. The idea of *rejecting the very question,* "Can both accounts of origins be correct?" is curious, but seemed to be popular with several in the audience at the Sokal-Ross debate. There are many things to note about the strategy of rejecting the question that have to do with the democratization of science.

For one thing, it is different from the silly relativistic response that *both* are correct, though it may have the same consequence. The motivation–a desire to empower the underdog–is perhaps the same in both cases, but a refusal to allow the political legitimacy of the question itself is an innovation.

Second, the objection assumes that Native Americans have but one view on this matter. In fact, there is a huge spectrum of views, just as there is a variety of opinions on most subjects among nonnatives. Many Native Americans cheerfully accept the standard migration account, and this, I think, presents a major problem. It's very easy to pass from "Let them believe what they want, if it empowers them" to "You're a traitor if you don't take a stand with the rest." A particular view can serve as a battle cry and a particular issue can focus attention. That, I suspect, is how Native American cosmology tends to function currently in political situations. The really crucial question, however, is this: What is the best way to promote Native American political aims? Would this way of democratizing science serve any but the narrowest and most temporary of goals, and would it even serve them?

We might also note in passing the general reluctance of postmoderns to support Christian fundamentalists who adopt the literal Genesis account of human origins. The double standard is glaring.

People's Science

The radical Science for the People group (with chapters based in Boston and elsewhere) was very active, especially during the Viet Nam war years.[4] It consisted of working scientists and others, all with a deep interest in the relations of science to the social order and the aim of making science better serve the oppressed. Among other things they are critical of current funding patterns, which tend to serve the ruling classes: There is, for example, a great deal of money for the study of heart disease or cancer, but relatively little for illnesses associated with malnutrition. There are studies galore on the behavior of the poor (useful for keeping them in line), but next to nothing on how corporate life really works (see Zimmerman et al., 1980).

The Science for the People group is also critical of "science for its own sake," because discovery and application are intimately linked. Although they admire the odd scientist who has refused to work in some field because the applications from that area of science have been pernicious, they nevertheless want something stronger and more effective than individual conscience. Their prescription calls for a large number of positive actions on the part of scientific workers, including:

- *Providing technical assistance to oppressed people.* This could include providing free medical clinics, expert advice concerning local environmental problems, or relevant research for third-world countries (i.e., on sugarcane for Cuba).

- *Researching the existing social structure.* Social scientists especially should work on exposing how power relations work and are maintained.
- *Demystifying science.* Information about science and technology needs to be disseminated widely so that ordinary people can gain greater control over their lives, and people must be given power over scientific decision-making when their interests are affected by those decisions.

These few proposals should give a feel for the general outlook of the Science for the People group.

There are two especially noteworthy features of the Science for the People proposals. The first is that they are not waiting for a democratic society so that we can then have a more democratic science–rather, science activism now is going to contribute to a better, more democratic society later. The second is their implicit assumption that science can help. There is such a thing as useful, objective science; there are facts about the human body and facts about how the social structure works such that if the oppressed knew them, they would have a powerful tool for improving their lives.

The only problem is the vagueness of the proposal for the demystification of science. It may, in fact, be hopelessly utopian to think that everyone can be well informed about the science that affects his or her life. What is needed is science *for* the people, not science *by* the people.

Understanding Science vs. Understanding Social Relations of Science

In *The Golem: What Everyone Should Know about Science,* Harry Collins and Trevor Pinch (1993) are quite concerned about scientific and technological decision-making in a democratic society. They sharply distinguish understanding the content of scientific theories from understanding the social relations of scientific experts. The former they say is not important for ordinary citizens–everything hangs on the public understanding of the latter (Collins and Pinch 1993, esp. 144ff). Though I have little enthusiasm for their social constructivist account of the content of science, I find their views on the democratization of science quite plausible.

The reason for rejecting as irrelevant the public understanding of the *content* of science is that technical issues are often very hard to understand and even well-trained experts often disagree on what is correct and what is incorrect. In these circumstances how can the general public be expected to do better? Collins and Pinch are, as I'll argue below, entirely justified in their argument.

Instead, Collins and Pinch want the public to understand how science works as an institution. We the public cannot make better guesses than the experts, but we can make informed decisions about which experts to believe. Expert testimony in legal cases is used as an analogy. A jury is often confronted with the testimony of rival experts–whose testimony should they accept? They are in no position to rethink the underlying theories, but they can note that, for example, one expert witness is employed by a company with an interest in the outcome; perhaps the other is a paid, itinerant expert witness who makes a living testifying in various law suits. This kind of information, they say, would be relevant in deciding whom to believe.

This special case is seen by Collins and Pinch as characterizing science in general. How could we make science more democratic? Not by increasing the public's knowledge of various scientific theories, but rather by informing the public about the various social relations that scientists enter into. Presumably this would be accomplished by teaching ever more science studies courses, getting the public to read more sociology of science, and so on. If I were to disagree with Collins and Pinch, it would be over details, not their premise, but the details are crucial to the plausibility of their argument.

First, it must be said that there are other steps toward the democratization of science than merely pointing out various social factors (though to be fair, Collins and Pinch do not claim any sort of exhaustive account). Knowledge of elementary statistics, for instance, can arm people against bogus arguments that seem quite persuasive. Here is a simple example: Canadians earn an average of around $2,000 a year less than Americans. Many Canadian politicians and business leaders argue that the average Canadian would be better off if Canada adopted a policy of lower taxes and no social programs. With a basic knowledge of statistics, one might ask about both the mean (average) and the median (middle) incomes in the two countries. The surprising fact is that the Canadian median is around $2,000 a year higher than the American median, even though the mean is about $2,000 lower. The apparent result (though this is still arguable, of course) is that the average (meaning typical) Canadian is financially better off with a policy of high taxes and strong social programs. Although many important social policy questions rest on complex scientific knowledge, this one can be grasped after a ten-minute explanation.

Second, there will be massive disagreement about what are and what are not existing relevant social relations. I often sharply disagree with most social constructivists over the causes of any particular scientific belief, so I will to the same extent disagree over the existing social relations that the general public should be aware of. In many cases (e.g., this chemist who is offering expert advice works for the company that stands to profit), the potentially relevant information is obvious. In so many other interesting cases, however, it is far from clear what is potentially of concern. We could even fall into a nasty regress. Should we believe expert *A* or expert *B*? Learning about their respective social relations doesn't help, because social historians α and β offer rival views about the social interests of *A* and of *B*. So now we must ask: Should we believe α or β? Alas, once again we could be faced with rival accounts of α's and of β's rival theories of social relations, this time coming from other social scientists α′ and β′. This process is potentially endless. The fear of regress might be thought a silly concern, but it's not. Determining social relations is as difficult as anything in physics or biology, and just as open to rival accounts, if not more so. One need only sample some of the bitter exchanges between Latour and Bloor, for example, to see how extensive the differences of opinion can be.

For many years now the scientific community has been worried about accounting for social relations. A recent issue of the *Journal of the American Medical Association* focuses on issues of peer review. One of the contributions argues for full disclosure of any sort of financial interest that the author of a scientific paper might have. It takes the view that "full disclosure [of financial interests] removes the suspicion that something of relevance to objectivity is being hidden and allows readers to form their own opinions on whether a conflict of interest exists and what relevance that has to the study . . . [and] believe that the scientific community and the public will be best served by open publication of financial disclosures for readers and reviewers to evaluate" (Krimsky and Rothenberg 1998, 225).

The only caveat I would add is that there are many other types of interests besides financial ones. Social constructivists, too, would stress this. But such interests are not easily disclosed. Would we ask Pasteur, for example, when he published his work against spontaneous generation to disclose his reactionary Catholic and monarchist political sentiments? The rich spectrum of social factors that constructivists have in mind goes well beyond immediate financial gain. Nevertheless, even financial disclosure would be a big step forward in the democratization of science, a step that Collins and Pinch would cheerfully endorse—and so would I.

Participatory vs. Representative Democracy

One thing that is stressed by almost all champions of a more democratic science is the capacity of ordinary people to understand scientific issues. (Collins and Pinch are the exception.) This is in part to combat "the cult of the expert," so often a mere camouflage for pursuits contrary to the public good. Yet there's a terrible ambiguity in the idea that the general public can understand scientific issues.

To make the point, let me brag about my linguistic abilities. I can speak several languages: French, German, Russian, Greek (both ancient and modern), Swahili, Chinese, Urdu, Cree, and a great many others besides my native English. Alas, what I *can* speak and what I *do* speak are distinctly different. I stumble through French and can barely ask street directions in German or Italian; my Chinese is limited to "hello" and "thanks." So why do I say I *can* speak all these different languages? Because (on the evidence of speaking English) I seem to have the usual brain structure, normal hearing, and standard equipment in the mouth and throat; in short, I have what's required to speak *any* human language, and in that sense I *can* speak them. But if you wish to discuss the fine points of an Aeschylus translation, I'll send you to the Classics Department; if you urgently want the Hungarian for "Where's the toilet?" I can only wish you the best of luck.

Ordinary people are capable of complete scientific understanding in the same way I'm capable of speaking many languages. They have the raw ability, and they have it at a sufficient level. We're kidding ourselves, however, if we think that ordinary people have the background, the resources, and the time to sort through complex scientific issues, even though most are capable of doctoral-level work in any science. Am I overstating things? Perhaps, but the point is that what people are in principle capable of and what they actually have the time and resources to accomplish are quite different. Ability isn't the issue, since even Nobel Prize–winners in physics and biology usually have only a superficial knowledge of each others' fields. So let's not expect more from the rest of us. Does this mean that we must succumb to the cult of the expert?

Before trying to answer this question, a few words about democracy itself. There are two contending views: *direct* and *representational.* The ideal of direct democracy is that all should participate in every decision. Fifth-century Athens is the paradigm, where every citizen was involved in every decision. (Never mind for now the glaring fact that this included only free, adult males.) Since the demise of the small city-state this form of democracy has been wholly impractical, but current champions of direct democracy point out that the Internet makes maximal participation of every citizen perfectly feasible once again. We could have referenda on everything. But feasible or not, why would we even want a direct democracy? Many reasons could be given: It's intrinsically good to be involved in the affairs of our society; it has a leveling effect that helps to prevent any monopolization of power; people become better citizens by becoming involved; and so on.

The alternative to direct democracy is representational democracy. Most current democratic governments function this way: Americans elect congressmen and senators to the Congress, Canadians and the British elect MPs to their respective parliaments, Russians elect deputies to the Duma. Representatives, it is often feared, can be stupid and lazy, or pressured by party politics, or even corrupted by the bribes of lobbyists. These fears are well founded, yet representational democracy has one great advantage over direct democracy, an advantage that may be overwhelming–*expertise.*

I'll illustrate with capital punishment. There's something of a paradox here, however. Any example (to be effective) must yield different outcomes for direct and representational approaches; yet, by its very nature the majority view (in the example) must be at odds with the representational outcome, so it will be an uphill battle to make the representational view look superior. Nevertheless, the point of any example should be clear, even if the example itself fails to persuade.

Opinion polls in Canada have for a very long time consistently favored a return to capital punishment. If we had direct democracy (say, in the form of a referendum on the issue), then we would very quickly start hanging people once again. In spite of the strong public sentiment in its favor, capital punishment has consistently been rejected by the elected representatives in parliament. Even though it is regularly an election issue with the winning campaigners often promising to support a return to hanging, it is still roundly defeated in the periodic "free vote" taken to decide the issue. Why?

I think the answer is rather easy to understand. Many elected representatives intend to support a return to hanging when first elected, but while in office, they learn that capital punishment is undesirable. At the outset they probably share their constituents' views that the murder rate is growing, that hanging will deter murder, and that it's cheaper to execute than to keep someone in jail for life. Once they start to examine the issues, however, elected officials find that the murder rate has actually declined since the abolition of capital punishment almost four decades ago. They also know that their constituents would want extensive safeguards in place to prevent the hanging of anyone innocent, but they find that this would be tremendously expensive. They learn that the Americans, for example, spend about 2 million dollars to execute someone, whereas it only costs about half a million to incarcerate for life. After learning these sorts of facts MPs change their minds.

With some justice, representatives who come to oppose capital punishment could still say they are carrying out the wishes of those who voted for them. After all, ordinary people who want capital punishment also want many other things. They want innocent people spared, they want murder deterred, but they don't want to pay unnecessary taxes or incur unnecessary brutality. Their representative comes to realize that all these wishes cannot be simultaneously satisfied. Capital punishment is simply incompatible with the rest of the wants of most ordinary people.

Any ordinary person could investigate capital punishment and most likely would come to the same conclusion. The problem is that it takes a fair amount of work. Time and effort are needed to figure things out—often more time and effort than ordinary people have to give. Representative democracy has this huge advantage over direct democracy—it opens the door to expertise acting in our interests.

Direct democracy stands in the way of knowledgeable and efficient thinking. It's open to abuse and even to parody. In the Canadian federal election of November 2000 one of the parties, the Canadian Alliance, urged referenda on any topic the public wanted. Cynics held that this right-wing party expected to be able to manipulate the public into supporting its favorite aims, such as reintroducing capital punishment, prohibiting abortions, prohibiting gay marriages, lowering taxes, and so on. The suggested mechanism said that a referendum would be held on any issue if 3 percent of the voting public signed a petition in its support. A television show that specializes in political comedy, "This Hour Has Twenty-two Minutes," launched a petition on the Internet demanding that the Canadian Alliance leader, Stockwell Day, be forced to change his first name to "Doris." They surpassed 3 percent support within hours, putting the idea of public referenda in complete disrepute. (To be fair, it should be noted that Canadians were already wary, given the bitterness of recent referenda on the proposed separation of Quebec.)

The example of capital punishment is almost trivial. Discovering its ins and outs is rather easy compared with, say, assessing complex statistical arguments used in race and IQ studies. If democracy is better served by the representative version when it comes to ordinary political matters, consider how much more a representative version is needed when we ponder complex scientific issues. Of course, ordinary people can understand these issues: they can see the pros and cons of regression analysis; they can see the horrible problems of divergent integrals in quantum theories of gravity and the amazing virtues of string theory in avoiding them; they can easily grasp the ups and downs of the adaptationist program in evolutionary approaches to human behavior. But they can understand each of these in the same sense that they can speak numerous languages. With the proper training any normal human can speak any language. Most people could achieve a high-level understanding of any branch of science, but only if several years have been devoted to its intense study.

Of course, there are degrees of comprehension. Understanding Einstein's general theory of relativity is not an all-or-nothing matter—neither is speaking Dutch. With time and effort, one's ability grows. Some branches of science, especially the nonmathematical ones, are easier to grasp than others. Yet let's not, in the name of democracy, fall into thinking that the general public is fully capable of making informed choices on scientific matters. Although it stems from a laudable, egalitarian motivation, it smacks of anti-intellectualism and it is unlikely to be helpful.

When we consider the democratization of the practice of science, let's do so in terms of representative democracy. To see how, let me begin by sketching a view of the growth of knowledge.

Comparative vs. Absolute

An earlier view of the growth of objective knowledge said that we take our theories to nature and test them against the evidence. According to this *absolute* view, which is now largely discarded, we conjecture a theory *T;* we draw out some of its testable consequences; we check to see if these testable consequences are true or false; we then accept those theories that have only true consequences (as far as we know) and reject any with false ones. Moreover, we also try to systematize the observable realm with our theories, that is, organize and explain diverse phenomena by means of a small number of powerful principles. This simple picture applies to the natural sciences, the social sciences, and indeed (with appropriate modifications) to any rational intellectual activity including literary criticism and even theology (so-called natural theology, not faith-based religion).

The crucial thing to note is that on this account of knowledge, a single theory can be evaluated *independently* of all other theories. Our conjectures are confirmed (to some degree) or refuted by the evidence—nature says yes or nature says no (though often nature speaks in a whisper). In consequence, it becomes utterly irrelevant that the person conjecturing may be the worst dogmatist and filled with the most vile prejudices, since the evidence supports or refutes the theory regardless of the theory's odious origin. (Philosophers refer to this as the well-entrenched distinction between *discovery* and *justification,* which I discussed in the third chapter.) The result is that we needn't worry about the genesis of a theory, because honest testing will reveal its true worth.

This is the *absolute* view of the growth of knowledge and there is much to commend it. However, considerable historical labor over the past several years has led to quite a different outlook. Since the work of Kuhn and others, rational theory-choice is seen as *comparative.* No longer do we think that theories can be tested solely by the evidence. Rather, theories can only be evaluated with respect to their rivals. Given some body of evidence, we can say that T_1 is a better theory than T_2; but we cannot say that T_1 is true unless T_1 is chosen from a more or less exhaustive pool of candidates.

Not only are theories evaluated by means of the evidence relative to their rivals, but what counts as evidence may depend heavily on what rival theories are being considered. Perhaps the best illustration of this is Paul Feyerabend's famous example of Brownian motion (1962), an example I described in detail in Chapter 4 and summarize below.

Early in the nineteenth century, a Scottish botanist, Thomas Brown, noticed a remarkable phenomenon. Tiny bits of pollen moved randomly around in a fluid. Were they alive? What was their source of energy? What is the cause of this bizarre motion? No one expected classical thermodynamics to explain this, any more than it could or should explain the color of snow. A rival to classical thermodynamics, the kinetic theory of heat claims that heat is average kinetic energy, the energy of motion of the tiny bits of matter. At the turn of the twentieth century, Einstein, Perrin, and others offered an explanation of Brownian motion: the relatively large bits of pollen are being knocked around by the tiny, rapidly moving molecules. The success of this explanation for the rival kinetic theory put classical thermodynamics on the spot. Suddenly, the phenomenon of Brownian motion was put into its domain; it was now obliged to explain it. Failing to do so was a strike against the theory. The moral is simply this: Brownian motion was relevant evidence for classical thermodynamics *only* because a rival had explained it. Without the rival, Brownian motion would have remained evidentially irrelevant.

The notion of absolute theory evaluation is held by many who think they are upholding the objectivity of science. With its rejection, many others abandon any form of objectivity at all. Both are mistaken. Comparative theory evaluation is, or at least can be, a perfectly objective and rational affair. This cannot be over-stressed. There is no question of people merely choosing on the basis of whim, or personal interest, or theory-based prejudice–though perhaps some do. The evidence forces (or at least inclines) us to a rational, objective choice, but the choice that the available evidence forces us to is the best among the *available* rival theories.

So now we face a big problem: What happens if the set of rival theories is skewed in some systematic way? How do we ensure that the set of rivals is the best possible? This turns out to be a fundamental question in epistemology; and it may have (at least in part) a political answer.

Getting the Right Representatives

Consider a rather stunning remark made by the great anthropologist, Claude Lévi-Strauss. He described a village as "deserted" when all the adult males had left. I doubt that any female anthropologist in the same circumstances would have said: "The entire village left the next day in about 30 canoes, leaving us alone with the women and the children in the abandoned houses" (quoted in Eichler and Lapoint 1985, 11). Of course, we're sensitized to avoid such statements today, but the problem is more than mere offensiveness to women. The passage reveals, I suspect, a certain view about the structure and operation of society and how best to explain it. The implicit principle at work with Lévi-Strauss is: "Look for the underlying social relations among adult males; when you've found those, you've found the key to understanding how a society works." Of course, Lévi-Strauss never expressed things this way; I doubt that he was even aware of making such a deep assumption. A female anthropologist in Lévi-Strauss's day may not even have challenged this viewpoint–a viewpoint that may, in fact, be an appropriate strategy for studying some societies. I dare say a female anthropologist–even one trained by Lévi-Strauss–might have conceptualized things differently on her own—not because she is bias-free, but because she has *different* prejudices.

We are all full of unwarranted background assumptions and inclinations of which we are more or less unaware. This is not a question of intellectual honesty. We do not know precisely what our biases are, so they cannot be systematically eliminated. We can do the next best thing: We can organize the pursuit of knowledge so that a great variety of *different* prejudices are at work in the production of theories. We can then select the best theory from among the rivals. The way to ensure the optimal diversity of rival theories is to make sure we have a wide variety of theorists. Currently, the pool of those who make conjectures is heavily skewed toward white males from upper-middle-class backgrounds living in wealthy countries. Future hiring must change the proportions so that a larger number of females, minorities, and others with different biases are included in the group of theorists. In short, affirmative action is needed for the sake of improving the growth of knowledge; pluralism for the sake of epistemology.[5]

Hiring into the scientific community is based on "merit"—at least ideally. This merit is construed in a rather narrow sense—it takes into account the intrinsic abilities of a candidate, but ignores the candidate's relation to the rest of the scientific community. If knowledge grows through comparative theory assessment, then being a "nonstandard" theorizer is a definite asset. Of course, a hiring committee that is racist and sexist is a poor one, but a committee that prides itself on being "gender- and color-blind" may not be that much better. For the sake of better science any hiring committee should seek to promote the sort of pluralism essential for optimal scientific progress. Very often, unfortunately, it is the larger society that must impose "affirmative action" policies on the scientific community. This is a very political act, but one that will result, paradoxically, in more objective science.

In sum, the way to democratize the practice of science is by having *representative* scientists from all relevant groups that have potentially competing interests. It's hopelessly utopian to want all people to understand all aspects of current science, but it's not unrealistic to want every branch of science to have participants drawn from every affected social group.

Is That All There Is to Democracy in Science?

Definitely not. There is a multitude of diverse respects in which science might become more democratic.

- *The practice of science.* Research could be carried out by a more democratically selected group of researchers.
- *The practice of reporting scientific results.* Disclosing one's social situation (especially economic) could be made part of the publishing process.
- *The fruits of science.* The benefits of scientific and technological research could be shared more equitably.
- *The glorious entertainer.* Sitcoms and football could give way (at least a little) to more highbrow literature and science, and access to these could be made more widely available.

The list could go on indefinitely. I've really only talked about the first sense–the practice of science–in which science might become more democratic. This is the principal one in the science wars, but others should be stressed, too. When it comes to the fruits of science, we need to ask if any class or group is benefiting at the expense of others. Is AIDS research cutting unfairly into the funding for breast cancer? Perhaps. Is the third world suffering from international patent laws in which large pharmaceutical companies withhold essential drugs? You bet. But how does the organization of noncorporate research help or hinder this grotesque situation?

What of the sheer pleasure of satisfying our intellectual curiosity? We sympathize with those who wish they had had the opportunity to study music. Those who can't play the piano have some idea of what they are missing. On the other hand, hardly anyone who is ignorant of it laments not learning calculus, but I believe they should, since they are missing out on an unparalleled pleasure. (I'm often seen as a hopeless crank for thinking this.) Making the delights of science available to all is yet another aspect of the democratization of knowledge. The glories of science contribute enormously to the uplifting of civilization itself, and I don't mind seeming a cornball for saying so.

9
Science with a Social Agenda

Bryan on Evolution

History has been unkind to William Jennings Bryan, characterizing him as a Bible-thumping buffoon. Bryan was anything but the bumpkin he is made out to be in common accounts of the Scopes trial. This infamous Tennessee court case in 1925 pitted Bryan against Clarence Darrow over the teaching of Darwinian evolution in the public schools. Darrow was probably America's greatest trial lawyer and a very prominent defender of progressive causes. In a famous Broadway play and later an even more famous movie of the event, *Inherit the Wind,* Bryan is a bumbling ignoramus in the grip of unthinking Christianity, a dogmatic dimwit, easily bested by Darrow. But Bryan was a more intelligent, more engaging, and more important character than that, and he is especially interesting for would-be science warriors today.

William Jennings Bryan was a great American hero, someone in whom people on the Left can take a sympathetic interest. A century ago, he was a leading populist politician, running three times for the Democratic Party as their presidential candidate. He was a strong opponent of the runaway capitalism of his day and fought hard and effectively for progressive social change, including votes for women, progressive income tax, and for getting the United States off the gold standard, which was a terrible burden on the American working class. He resigned as Secretary of State in 1915 when Woodrow Wilson decided to enter the Great War. Because of Bryan's very strong support for progressive causes, Darrow, a fellow left-wing Democrat, campaigned hard for him when Bryan sought political office.

Bryan was also a devout Christian, but not so literal-minded that his opposition to Darwinian evolution was inevitable. His anti-evolution crusade stemmed instead from his social activism. Bryan was appalled by the wretched social conditions endured by the workers of his day, conditions that were supposedly justified by Carnegie, Rockefeller, and others on Darwinian grounds. Social Darwinism, as it's often called, was used to exonerate winner-take-all capitalism. If many fall by the way, it's perfectly appropriate, since life is a struggle for survival and only the fittest would or should endure. Rectifying the situation was tampering with the natural order, a sin against nature itself. Bryan rightly despised this theory.

Besides pernicious capitalism, the eugenics movement and the Great War were, according to Bryan, tied to Darwin's legacy. Darwin's son, Leonard Darwin, was president of the Eugenics Education Society, and Ronald Fisher, the brilliant statistician and evolutionary biologist, was also a keen eugenicist. As for the Great War, there were numerous reliable reports giving firsthand accounts of the ideology of the German High Command. They claimed that their aggressive actions were justified by the fact that life is a struggle in which only the fittest will survive. Bryan was horrified.

Bryan's Christianity was not liberal, nor was it sophisticated, but if we look for the prime source of his opposition to evolution, we will find it in his concerns for social justice. Religion merely shaped the form of his opposition. There were in addition important democratic considerations involved. The vast majority of American voters, thought Bryan, believed in the Genesis account of human origins, and democracy demands that we teach what the people–rather than an educated elite–want taught in the schools they pay for with their taxes. In short, Bryan cared about who should rule.

Though Bryan's anti-evolution crusade was a campaign both for social justice and for democracy, the same cannot be said of his anti-evolution allies. William Bell Riley, a prominent fundamentalist leader, attacked the "Jewish-Bolshevik-Darwinist conspiracy"[1] and praised Adolf Hitler for doing his part in suppressing it. The Ku Klux Klan stood firmly for God and against Darwin. For the most part, activist opponents of evolution were a disreputable lot. Bryan was not of their ilk.

In the famous courtroom confrontation between Darrow and Bryan, the latter was shown to be hopelessly ignorant of evolutionary theory. Science was not his strong suit. Darrow exposed more than Bryan's scientific ignorance, however. It turned out that Bryan was also shaky in his theological convictions. When pressed on biblical details, he wavered. How, for example, could there be day and night *before* God created the Sun? Bryan seemed not so literal in his reading of the Bible after all, taking "day" to mean "age," which could be quite long. (This lost him considerable fundamentalist support.) Bryan came off looking like an inconsistent and dogmatic dimwit, which is a great pity, since social justice, not theology, was the main reason he got into this mess. His case against Darwin stemmed from the highest motives. He saw evolution (as it was often presented in his day and is still sometimes presented in ours) as a doctrine that justifies *might makes right,* and he was perfectly justified in objecting strenuously to it. But he was hopelessly ineffective. If there is a moral to be drawn, it is a simple one: Don't fight for social justice Bryan's way.

Gould on IQ

In the public realm no one has been as prominent and influential as Stephen Jay Gould in debunking myths surrounding biology. Whether it be social Darwinism, or sociobiology, or IQ, or any other form of biological determinism, Gould has taken up the cudgels to great effect. Like Bryan, Gould is appalled at some of the uses to which modern genetics and Darwinian evolution have been put. (He has even written sympathetically of Bryan.[2]) There is a world of difference in their two approaches, however. The motivations are the same: Both are on the Left and they reject the socially pernicious biological determinism that is an alleged consequence of various biological theories. The difference is this: Bryan rejected biology (and upheld the Bible in its place). Gould rejects the claim that social Darwinism follows from Darwinian evolution or from modern genetics. Both encourage science with a social agenda (or in Bryan's case, antiscience with a social agenda), but the contrast couldn't be more striking–nor could the effectiveness.

What does Gould care about? Like anyone on the Left, he detests the class structure in American society and the racism that goes along with it. IQ studies play a role in this. Is intelligence something that is distributed randomly through the population? Is it inherited? Is it a product of the environment? Social conservatives often claim that those on the bottom of the social heap are there because of their inherent lack of worthiness. The social order reflects the true merit people possess. Add to this a doctrine that intelligence is inherited, and we easily arrive at the conclusion that there is nothing to be done for the unfortunate lower orders. They just can't cut it. Money spent on special programs, according to this view, would be wasted. The poor have not been discriminated against unfairly; they are located in the social stratum exactly where they *naturally* belong. Race can be added to the mix. African Americans are disproportionately in the working class because, says this line of argument, they are inherently less intelligent than whites.

When you're on top, there is nothing so gratifying as scientific evidence that you deserve to be there. This is the message of *The Bell Curve,* the massive work by Herrnstein and Murray (1994). Gould remarks that the success of *The Bell Curve* "in winning attention must reflect the depressing temper of our times–a historical moment of unprecedented ungenerosity, when a mood for slashing social programs can be powerfully abetted by an argument that beneficiaries cannot be helped, owing to inborn cognitive limits expressed as low IQ" (1994, 11). In short, this is science with an agenda, a very conservative (even reactionary) one. Gould, of course, has the opposite social aim and his attack is very effective.

Gould notes that the argument of *The Bell Curve* rests on a number of shaky assumptions. For instance, it posits that there must be a single thing, usually called the *g*-factor for general intelligence. It doesn't accept the notion of multiple intelligences (say, a distinct intelligence for music or for math or for working with animals). Second, intelligence must be heritable. If it showed up at random in the population, then low IQ could not explain why working-class parents tend to have children who remain in the working class. IQ must also be immutable, because if it was malleable then it should be raised by appropriate early education programs, just as athletic prowess and musical performance are greatly enhanced by coaching and practice.

Heritability, it must be stressed, is not enough to show immutability. Herrnstein and Murray claim that when someone has a low IQ, that person is simply stuck with it. Consider height, however, which is obviously genetically determined. Look how it has increased in the past few decades. North Americans are considerably taller today than they were three generations ago. Why? Better nutrition. Perhaps in other conditions than those in which they find themselves, parents with low IQs today might produce offspring with increasingly higher IQs in the next few generations, even though there has been no genetic change. So, even granting Herrnstein and Murray the heritability of intelligence (but only for the sake of the argument), immutability still does not follow.

Gould further charges the authors of *The Bell Curve* with deliberately passing over strong, but indirect evidence that suggests a contrary conclusion. For example, the spread in IQ between black and white in the United States is about fifteen points. One would expect some of this difference to show up in the children of mixed unions. American soldiers stationed in Germany fathered many natural children who were raised as Germans by their German mothers. Yet, there is no discernable difference in the IQ scores between those with black and those with white fathers.

This is just a sample of the types of criticisms Gould and many others have made of IQ studies.[3] This head-on approach is much more effective than Bryan's approach. Instead of turning away from science altogether, Gould and others simply show that Herrnstein and Murray fail as scientists: they ignore available evidence, they make faulty inferences, they make questionable assumptions. This is not an indictment of science; it is an indictment of two bad scientists.

Education and Democracy

How might Bryan handle a case like this? He was a staunch majoritarian. If the majority wants the biblical account of creation taught in their schools, then that is exactly what they should get. By a parallel argument, if people thought that research connecting IQ to race or class was socially destructive and not worthy of support by their taxes or exposure in their schools, then Bryan's majoritarianism would compel him to want it banished, as well. To let a scientific minority determine the content of the school curriculum over the wishes of a majority is simply undemocratic. As he put it: "A scientific soviet is attempting to dictate what shall be taught in our schools" (quoted in Larson 1997, 135). For Bryan this was a matter of deep political principle. Even if scientists are correct about evolution (or correct about the connection between class and IQ) and the people are wrong in their beliefs, still the majority rules–not the scientific elite. Science, Bryan insisted, must serve the people–not the other way around.[4]

Though hardly any thinking person would today support Bryan on evolution, I dare say many would be tempted to a similar view when it comes to IQ studies. It is, I think, an important dilemma: Who should rule? A simplistic democratic solution is tempting, but should be avoided.

The trouble with Bryan's democratic position is that no majority opinion will be stable if it flies in the face of an overwhelming scientific consensus. Those who stick to Genesis and reject Darwin quickly find themselves having to reject much of contemporary geology (which posits a 4-billion-year-old Earth), some physics (involving radioactive dating), most cosmology (which holds that the universe started 15 billion years ago in the Big Bang), a great deal of historical linguistics (a far cry from the "Tower of Babel" account of the origins of different languages), and so on. We can admire the stout hearts of those who stand firm against all of this–but not their mushy minds.

A second–still democratic–solution is this: The people rule, but they appoint a body of scientists to advise them, and even give them enough power to overrule the opinion of the majority–with the majority's indulgence. We see this in action when an elected government sets up, say, a body of experts to oversee the production of food and drugs. That body can prohibit certain products, even when the majority holds a contrary view of the matter. The power given to the regulating body can be removed by majority rule, which is what keeps the process democratic. We obey our "doctor's orders" not because we are genuinely obliged, but because we choose to obey them. In the case of evolution, the scientific consensus is so strong that we rightly include the theory in our school curriculum and keep the Genesis account out. There is no consensus one way or the other in the case of IQ studies, so we may choose to omit it from the standard school program and keep it from having a pernicious influence on social policy. When we consider its social divisiveness, it's obviously proper to omit it.

The second democratic solution still falls short of the whole story. It ignores the social content of the science itself. Bryan was motivated to reject Darwin because he detested what he took to be the social consequences of evolution, a hateful justification of the doctrine that might makes right. Gould has the same motivation, but instead of an across-the-board rejection of evolution, he shows that certain socially destructive consequences do not follow from the theory, contrary to what some (though not all) sociobiologists might say. In race and IQ studies, he shows that the nasty conclusions drawn by Herrnstein and Murray and others are simply unwarranted–unwarranted by the lights of orthodox science. Criticism, however, could run much deeper than these examples would suggest. In previous chapters we saw that straightforward observation and inference will not lead invariably to the truth. Sometimes the problem with a theory cannot be seen except in the light of serious alternatives. This is where my proposal from the last chapter about the democratization of science comes into play. In the next few sections we'll look at examples.

Feminist Critiques

One of the most infamous expressions in all of science studies is "Newton's rape manual." Sandra Harding coined the phrase–and has been paying for it ever since. It has been quoted repeatedly as decisive evidence of the insanity of feminist critiques of science. Some "evidence" might be less than it seems, however, so let's take a look at the context of her offending remark.

Harding is interested in the role of metaphor in science. It's one of many topics in *The Science Question in Feminism* (1986). In it she notices a tension in the views of several philosophers and historians of science. On the one hand, many have stressed the great importance of guiding metaphors: they have played and continue to play a key role in the course of research and in shaping our general view of nature. Taking nature to be a great machine (as was common in the seventeenth century) is one of the most striking examples.

There have been other metaphors than the mechanical ones, however. Feminist historians have repeatedly cited sexual examples such as those of Francis Bacon, who represents the relation between scientist and nature as a man dominating a woman. Here are a few representative passages from Bacon: "Let us establish a chaste and lawful marriage between Mind and Nature . . . [N]ature betrays her secrets more fully when in the grip and under the pressure of art than when in enjoyment of her natural liberty . . . [We can expect much from] the nuptial couch for the mind and the universe . . . I come in very truth leading to you Nature with all her children to bind her to your service and make her your slave" (all cited in Lloyd 1984, 10–17). When feminist historians claim that these domination metaphors have played a role harmful to women (and to the environment), they are rebuffed with the remark, "mere metaphors"; but the inconsistency is rather glaring. If machine metaphors are effective, why not gender metaphors? "Presumably," Harding argues, "these metaphors, too, had fruitful pragmatic, methodological, and metaphysical consequences for science. In that case, why is it not as illuminating and honest to refer to Newton's laws as 'Newton's rape manual' as it is to call them 'Newton's mechanics?'" (1986, 113).

Harding's point is that we can't have it both ways. If metaphors are effective in science, then Bacon's sexual domination metaphors can't be passed off as harmless fluff. No one needs to be told about the dangers of quoting out of context. In this case, remarkably little context is needed. Three paragraphs are sufficient to make Harding's point perfectly clear and to see the injustice of the charge commonly raised against her. (The frequent charge of shoddy scholarship made against feminists is particularly ironic.)

The feminist spectrum is very wide. One end of it is repeatedly cited and debunked. The debunking of particular cases may be completely just, but the inference to the whole is without warrant. Frequent targets are Luce Irigaray and Kathryn Hayles and their work on fluid mechanics. Hayles writes,

> The privileging of solid over fluid mechanics, and indeed the inability of science to deal with turbulent flow at all, she [Irigaray] attributes to the association of fluidity with femininity. Whereas men have sex organs that protrude and become rigid, women have openings that leak menstrual blood and vaginal fluids. Al-

> though men, too, flow on occasion—when semen is emitted, for example—this aspect of their sexuality is not emphasized. It is the rigidity of the male organ that counts, not its complicity in fluid flow. These idealizations are reinscribed in mathematics, which conceives of fluids as laminated planes and other modified solid forms. In the same way that women are erased within masculinist theories and language, existing only as not-men, so fluids have been erased from science, existing only as not-solids. From this perspective it is no wonder that science has not been able to arrive at a successful model for turbulence. The problem of turbulent flow cannot be solved because the conceptions of fluids (and of women) have been formulated so as necessarily to leave unarticulated remainders. (Hayles 1992, 17)

It's hard to take things like this seriously. Sokal and Bricmont (1998) don't, and neither does Sullivan (1998), who examines this work in detail and finds many technical mistakes. The unfortunate thing is that criticisms of specific feminists' writings—which are perfectly just reproaches—too easily transform into rebukes of feminist approaches in general.

Feminists are not the only reformers, but they may well be the most systematic and comprehensive. Within the spectrum of feminist critics of science there is a large cohort who uphold rationality, objectivity, and scientific progress; but they also claim that science has been driven by value considerations that are far from ideal. Such critics of science differ from either the nihilists or the naturalists. Against the nihilists, this group of feminists embrace such traditional notions as objectivity, though they will certainly claim that it has been sadly lacking too often in the actual history of science. Against the naturalists, they see their role as more than mere chroniclers of actual science; they are out to make it better. To paraphrase Marx: the naturalists and nihilists have only interpreted science; the point is to improve it.

Values

Almost everyone acknowledges that there are values at work in science. The real debate is over their role and extent. For a start, we can all agree that values determine the research goals of science. A great deal of funding stems from hoped-for technological or military applications. Values certainly determine which fields or sub-fields get investigated, but do they play a role in the actual findings, that is, in the very substance and content of science? That's the big question. Scientific orthodoxy says no. A great deal of money goes into high-energy physics in hope (or in fear) that it may have a sizable payoff. But the reigning theory, known as the standard model, was adopted because of the evidence, not because the military paid for the work (whether American or Soviet). In medical research, AIDS activists have been responsible for an enormous growth of funds directed at combating AIDS, much to the chagrin of others who would like to see some of that money spent, say, on breast cancer research. Nevertheless, the finding that AIDS is caused by a certain kind of virus that affects the immune system seems completely independent of the politics of research funding. In short, values determine the *direction* of scientific research, but they do not, according to scientific orthodoxy, determine the *content* of the theories we ultimately come to believe.

I must immediately add a caveat or two. Though we can make a clean distinction conceptually, in practice the difference between direction and content is not easy to sort out. Increasingly, research funding comes from private concerns with a financial interest in the outcome, but the science wars are chiefly about the content of scientific beliefs, and that is what I will focus upon. Do values and interests play a role in determining the content? Are values, interests, and various political considerations present in the very heart of our scientific beliefs?

It's a pity that the debate is so focused. Much of the effort spent fighting the science wars would be better spent directly debating the sources of funding and the direction of research. Should proportionately more money go to AIDS or to breast cancer research? Value questions such as these can be discussed largely without having to settle the question of values in the content of rival theories. Nevertheless, even though there is a clear conceptual distinction between direction and content, the two issues are inevitably related. Pharmaceutical companies, for instance, are not likely to fund a line of research that won't be profitable to them. A research group's funding tomorrow can be linked to how "cooperative" they are today. The whole process borders on the scandalous. It would be a pleasant change to see those who rail against political correctness direct some of their energy to this real threat to academic freedom, not to mention human health. (I take this up briefly in the Afterword.)

We need to remind ourselves of the brief discussion of values back in Chapter 5. There we distinguished between cognitive values (such as: choose theories with greater scope and explanatory power) and other noncognitive values (such as: choose theories that are compatible with Christianity or choose theories that will regain the public's favor). We also need to recall that any example cited will be contentious and that the very distinction between cognitive and noncognitive values may turn out to be bogus.

In a pinch, we can imagine four rival attitudes toward the question of noncognitive values in the very content of science:

1. There are no noncognitive values at work in science.
2. All of science is impregnated with noncognitive value; there is no objective distinction to be made between good science and bad.
3. Noncognitive values sometimes play a role in science, but they invariably lead to bad science; good science is free of noncognitive values.
4. There is an important distinction to be made between good science and bad science, though both are shot through with noncognitive values.

The first of these is hopelessly naive. I don't know of anyone who believes it. The second is typical of that cluster of social constructivist views discussed much earlier under the headings of nihilism and naturalism. Interests and other noncognitive values are taken to be at work everywhere; there is no serious objectivity of any sort. Earlier in this book I discussed why these first two views are hopelessly implausible.

The third and fourth views are the really interesting ones, and are the main subject of this chapter. The third is close to scientific orthodoxy and allows that there has been some–perhaps even a lot–of bad science, usually brought on by some sort of intellectual corruption. Typically we hear that the Soviet biologist Lysenko (everyone's favorite example) produced bad science because he was in the grip of an ideology, whereas Newton, Darwin, and Einstein, on the other hand, produced value-free science.

Noam Chomsky exemplifies this attitude. When he criticized postmodern attacks on science and reason, he said that their complaints are sometimes justified, but only when reason had been perverted. "The critique of 'science' and 'rationality' has many merits," he says. "But as far as I can see, where valid and useful the critique is largely devoted to the perversion of the values of rational inquiry as they are 'wrongly used' in a particular institutional setting" (Chomsky, *Z-Net*). In short, there is such a thing as good scientific method, according to Chomsky, and it appears to be value-free. Values can pervert that method and lead to socially harmful results. Existing methods are fine as long as they are not corrupted by ideology. (Of course, even "good" science can lead to harmful results, e.g., nuclear physics and bomb-making.)

There is an account of the proper role for sociology of knowledge that goes along with this view of values in science and is connected to the so-called A-rationality principle, discussed earlier. We look for a sociological (or psychological) explanation of belief only when that belief is irrational. Bloor and others have ridiculed such an approach to science, calling it the sociology of error. The strong programme (and naturalism in general) is in complete opposition to this way of looking at intellectual life.

The fourth view of values goes well beyond the third and is characteristic of a great deal of recent feminist philosophy of science. On the one hand, it stresses that there are noncognitive values galore in the sciences, but on the other hand, it wants very much to avoid falling into a nihilistic and self-defeating relativism. On this view, there is still a genuine objectivity in scientific decision-making–some theories really are scientifically better than others.

One of the most important but underappreciated aspects of feminist critiques of science is the normative aspect. Feminist philosophers of science are out to change science for the better. Bloor and other naturalists are in the business of *describing* how science works; they are not in the business of *prescribing* changes. The same can be said for the anthropologists in the lab who–like most anthropologists–want to interfere as little as possible with the society they study. Of course, how could they adopt a different attitude? They don't acknowledge an objective difference between good and bad science (not in a normative sense), so they could hardly advocate one theory over another on the grounds that one is objectively better. This is the crucial difference between naturalists, on the one hand, and most feminist philosophers of science, on the other. Though all embrace elements of social constructivism, the naturalists are merely interpreting or describing what the scientists have constructed. The reformers are out to reconstruct it.

From the point of view of those trying to make science better, we will look in some detail at both the third and fourth positions, but will examine them through examples.

Man the Hunter

The search for human origins has enormous social ramifications. It informs our picture of ourselves–both anatomical and social–and so plays a role in the determination of social policy and civil life. One prominent hypothesis is the "man-the-hunter" view. The development of tools, on this account, is a direct result of hunting by males. When tools are used for killing animals and for threatening or even killing other humans, the canine tooth (which had played a major role in aggressive behavior) loses its importance, and so there will be evolutionary pressure favoring more effective molar functioning, for example. Thus human morphology is linked to male behavior. Male aggression (hunting) is linked to intelligence (tool making). On this account women play no role (or only a secondary one) in evolution. We are what we are today because of the activities of our male ancestors.

This is not the only view of our origin. A theory of more recent vintage is the "woman-the-gatherer" hypothesis. Some of the main contributors include: Sally Slocum (1975), Nancy Tanner and Adrienne Zihlman (1976), and Adrienne Zihlman (1978 and 1981). This account sees the development of tool use to be a function of female behavior. As humans moved from the plentiful forests to the less abundant grasslands, the need for gathering food over a wide territory increased. Moreover, women are always under greater stress than men since they need to feed both themselves and their young. Thus, there was greater selective pressure on females to be inventive; so tools were a result of female innovation. Why, on this account, should males lose their large canine teeth? The answer is sexual selection. Females preferred males who were more sociable, less prone to bare their fangs and to other displays of aggression. So, on the woman-the-gatherer account of our origins, our anatomical and social evolution is based on women's activities. We are what we are today largely because of the endeavors of our female ancestors.

The kinds of evidential consideration thought relevant in deciding this issue include: fossils, objects identified as tools, the behavior of contemporary primates, and the activities of contemporary gatherer-hunter peoples. Obviously, each of these is problematic. Artifacts are few and far between, and are little more than fragments; some tools such as sticks will not last the way stone tools will, so we may have a very misleading sample of primitive artifacts; moreover, it is often debatable whether any alleged tool was really used for hunting an animal or preparing it for eating, rather than used for preparing some vegetation for consumption; and finally, inferences from the behavior of contemporary primates and gatherer-hunter humans to the nature of our ancestors who lived 2 to 12 million years ago is a huge leap.

None of these considerations should be dismissed out of hand; each provides evidence of some sort, albeit weak evidence. We have a case of underdetermination–there simply isn't enough evidence to pick out a winner from these two rival theories. However, our ability or inability to find the correct theory is not important. The moral of this example is that it displays how value-laden background beliefs can affect scientific choices. If one is already inclined to think of males as the inventors of tools, then some chipped stone will be interpreted as, say, a tool for hunting. This will then become evidence in the man-the-hunter account of our origin. On the other hand, if one is a feminist then one might be inclined to see some alleged tool as an implement for the preparation of vegetable foods. On this interpretation the tool becomes evidence for the woman-the-gatherer account of our evolution.

We may hope, as Helen Longino notes, that "In time, a less gendercentric account of human evolution may eventually supersede both of these current contending stories" (1990, 111). It will not, however, be a value-free view of the matter. Values will always play a role in any scientific theorizing; it's just a matter of making the operative values visible. The great merit of the woman-the-gatherer theory is that its very existence made manifest the previously hidden assumptions of the prior man-the-hunter theory. Until the existence of the rival, the evidence for the man-the-hunter account was "dependent upon culturally embedded sexist assumptions" (1990, 111).

The moral to be drawn from this example is simple enough, though many-sided. Science with a bias needn't be bad. The man-the-hunter theory is not stupid. It's elaborate, thoughtful, and plausible. Its rival, the woman-the-gatherer theory, was the product of a different social outlook, but that doesn't make it silly either. It, too, is elaborate, thoughtful, and plausible. To its extra credit, it provided a relief for the older theory. Background assumptions and values became visible by contrast. There is even a next step that suggests itself. Are the categories man-the-hunter and woman-the-gatherer the correct ones? They arise in our current sexual division of labor, but are they justly imposed on the very distant past?

The example of rival origins theories is becoming a stock example for feminists. It nicely illustrates the point made in the last chapter about the democratization of science. Science is not only more democratic by including within its ranks a greater variety of theorists, but it also has the potential to produce better science.

There are other favorite feminist examples, such as primatology, which has been revolutionized by women with feminist interests. Yet another popular example, this time from reproductive biology, is particularly interesting. Textbook discussions of reproduction (at least in the past) typically described the sperm as active and the egg as passive. Suspicions that social stereotypes were being projected into reproductive biology are hard to resist when one encounters such metaphors as "Sleeping Beauty" and "Prince Charming" for egg and sperm, respectively. In the 1980s a different outlook was proposed (Schatten and Schatten 1983). Instead of complete passivity, the egg was given a major role in actively bringing about fertilization. It is surrounded by microvilli (tiny hair-like growths on the surface) that grasp and hold a nearby sperm; digestive enzymes in the sperm allowing it to enter the egg are activated by contact with the egg. Thus, far from being passive in this process, the egg is a very active partner in fertilization. The value of feminist input in this research has been to focus on the role of the microvilli, which were discovered about a century ago, but had not been seriously investigated until recently.

Examples such as this one concerning reproduction and the woman-the-gatherer example described above have been controversial. Even if they are rejected—and remember, most theories are rejected—they still illustrate a remarkable point. Background can make a huge difference in the style of theorizing. Once the rivals are on the table, we can perhaps make perfectly objective decisions about which is best, but the rivals have to be there for us to consider. Without feminists proposing new theories in the first place, we wouldn't even be aware of some of the possible candidates for the truth about reproduction or the truth about human origins.

Helen Longino hopes for a better account than either the man-the-hunter or the woman-the-gatherer hypotheses, since they are both too gender-oriented. Emily Martin (1991, 1992) worries that the new view of reproduction has its own set of problems. The egg, according to her, has been given a position of equality, but only by "masculinizing" it. Instead of a passive female, it is now, in effect, an aggressive male. One type of social virtue has given way to another. Rather than eliminating social stereotypes that were getting in the way of good science, we may have eliminated some only to replace them with others. As Londa Schiebinger might add (1999), the new account of egg and sperm too nicely mirrors currently favored views of gender relations—the two-career couple, working in a harmonious partnership. Although science continues to improve, social influences still require unearthing.

Doing science as a feminist—that is, theorizing from feminist motives—has proven highly effective. Conjecturing new theories that put women in a more favorable light and putting those theories to the test have been helpful. The strategy is not guaranteed to succeed, however, and I shouldn't be taken as saying that any political or social cause will succeed if it is wedded to objectivity. Quite the contrary. The clergy, for example, have suffered greatly at the hands of science. Occasionally they try to find support for religion inside science, but these attempts are usually laughable, the "anthropic principle" being the latest fad among graspers at straws. If one has a case, then objective science can bear it out, but only when one's view is well represented among the participating scientists. As long as women were shut out of science, then there was little hope of a woman-the-gatherer account of human origins being developed at all, and there was only a minute prospect that the possible active role played by the egg in fertilization would be taken seriously. Science with an agenda can be highly successful and utterly objective, but one's most ardent aspirations and earnest convictions may be dashed. Democracy is in the proposing–nature still does the disposing.

Doing Our Bit

I began this book with the Sokal affair and here is where I want to end. Sokal performed a great service; not in showing that the postmoderns and other social constructivists are hopelessly confused, but in opening up some elbow room for the pro-science Left, a group who have taken the back seat for far too long.

The science wars are fought on a many-sided battlefield. Some social constructivists are as pro-science as anyone could be–Bloor and his fellow naturalists, for example. Others are rightly thought to be hostile to the very idea. Enmity is widespread. Doubtless, there are many reasons for the animus, but the most important stem from social and political concerns. No one with a shred of interest in the well-being of others can be sanguine about the current social situation. In many countries there has been a tremendous shift in wealth from poor to rich, and the present plight of the impoverished is a scandal that must forever shame those who have milked the underprivileged to the point of despair.

Science should be the friend of the oppressed. Yet it must appear to many that science is part of the problem. We are regularly bombarded with "discoveries" of some gene or other that invariably seems to explain why the poor have little and the rich are doing so well. Can you imagine *Time* magazine running a cover story: "Scientists discover the socialism gene: 'It's the natural way to live' says Professor Big Shot of Ivy League U."? Fat chance. But there was a cover story on *The Bell Curve,* and it contributed, no doubt, to a widespread feeling that the current social order is perfectly natural, with the races and classes located just where they belong.

We often dismiss politically motivated social constructivists with the remark that clear thinking will be more useful than fuzzy nonsense. This is Sokal's view. I've often said such things myself (though not as dramatically as Sokal), and I'm happy to stand by them. But this is not enough. We can–and should–contribute to society's betterment. Scientists and especially philosophers of science are uniquely situated to rebut the pseudo-scientific justification of social inequality. More than anyone else, they have the skills to ferret out bad science, especially when it is serving socially pernicious ends.

The first task, of course, is to get things straight; alleged pseudo-science must be proved to be so. More than this is needed, however: The results must be publicized. Although no one has done as much as Stephen J. Gould to expose race and intelligence studies for the garbage that they often are, I wouldn't be completely surprised if that specific issue of *Time* magazine devoted to race, class, and IQ, which seems to linger forever in dentists' waiting rooms, might have a greater readership than all of Gould's numerous publications put together. Finding the problems with pseudo-science is one thing–bringing them effectively to the general public's attention is quite another. We should take every opportunity in general public forums to point out the problems. We should challenge our local psychologists, or economists, or biologists who are responsible for bad science to public debate. We should write popular articles for our local newspapers and international magazines, anything with a wide readership. One article in a major magazine will find a wider readership than a lifetime of publications in regular academic journals.

Who should rule? The people, of course. They just need to hear more intelligent and informed voices. If Sokal's hoax prods those who are analytically minded and sympathetic to science into socially constructive action, then that will be his true legacy.

Afterword

Like Wordsworth, "My heart leaps up when I behold a rainbow in the sky." There is much to be said about rainbows that isn't part of the usual domain of science. Rainbows were explained by Descartes long ago and his account has largely stood up. It has to do with the refraction of light in drops of water. There are lots of other accounts of rainbows that are just plain wrong. Among the false accounts is the one that says rainbows are a sign from God that he won't flood the earth again. Rainbows aren't that, but they are beautiful.

We often hear that there are many ways to understand reality, and that science is just one of these–none should be thought better or worse than others. To the liberal-minded, this is one of the big appeals of social constructivism. If this means that rainbows are as Descartes says *and* they are beautiful, then who could object? But claiming more than this is rubbish, if it also means that rainbows are something else entirely. It's one thing for a cultural group to have its own political goals, but quite another to have its own version of science.

In saying this, there is a great worry–dogmatism. Science is fallible and Descartes's theory might be wrong, but that does not mean that all perspectives are correct. It only means that some other theory–so far undiscovered–is the true one.

Who Rules in Science? has been a defense of science, though of science done a certain way. Social constructivists have been right to think there are loads of social factors at work, but they have been wrong to think that reason and evidence are not playing a dominant role. Scientific objectivity is both possible and actual. Values and objectivity can and do co-exist. Of all the things that could be called the science wars, the relation between epistemology and politics is arguably the most central and most important to understand. There are other issues, too, and some of them involve politics. We could discuss at great length a number of overlapping topics under the heading "science wars," but in this Afterword I'll only briefly mention a few of them.

One concerns religion and science. Religious and social conservatives are more often than not happy with science and technology so long as it has no bearing on cherished beliefs about religion and morals. Darwin, of course, is a lightning rod. There are endless battles over evolution, especially in the United States. Anti-evolutionists are often in high places. While he was president, Ronald Reagan said that evolution is "only a theory." His intent was to cast doubt on Darwin, but among scientists, it only cast doubt on Reagan's sanity. The religious Right has had remarkably little impact on science itself: its real threat is to science education. American school boards seem bent on making American children as poorly educated as possible.

Another concern that is part of the science wars is environmentalism. It is hard to imagine the environmental movement without standard science. Standard science discovered the hole in the ozone layer over the Antarctic and its connection to the increase in skin cancer. Standard science detected the presence of toxins in drinking water and the adverse effects they have on our health. Standard science created a multitude of models of the weather, each indicating potentially harmful global warming. The environmental movement is most effective when it acts on the information provided by standard science. Without that information, any action would be unmotivated. Yet some environmentalists have an ambivalent attitude toward science. Perhaps the picture of nature that contemporary science gives us is fundamentally flawed. Much of its success is tied to a reductionist and mechanistic attitude. This outlook leads us to think that nature can be divided into isolated parts which can be manipulated without consequence for the rest. Such an attitude, these critics contend, can lead us into thinking we can introduce a particular new crop in some part of the world without ramifications for other vegetation or for preferred ways of life.

Perhaps the greatest threat to science is not social constructivism, not the religious Right, and not the attitudes of environmentalists. It is the commercialization of knowledge. This issue is closely connected to the central themes of this book, since it involves both epistemology and politics, but it is also a topic that can be addressed directly and independently. It seems appropriate to close this book with a little tirade.[1]

In recent years, we have all watched the increasing commercialization of the campus. The numerous advertising posters and the golden arches of fast-food outlets may be an affront to our aesthetic sensibilities, but they are, arguably, no worse than ugly. Some of the other new features of commercialized campus life do, however, constitute a serious threat to things we rightly revere. Privatization and the business model are the potential menaces.

What do these notions mean? They involve an increased dependence on industry and philanthropy for operating the university; an increased amount of our resources being directed to applied or so-called practical subjects, both in teaching and in research; a proprietary treatment of research results, with the commercial interest in secrecy overriding the public's interest in free, shared knowledge; and an attempt to run the university more like a business that treats industry and students as clients and academics as service providers with something to sell. We pay increasing attention to the immediate needs and demands of our customers and, as the old saw goes, "the customer is always right."

Privatization is particularly frightening from the point of view of public well-being. A researcher employed by a university-affiliated hospital in Canada, working under contract with a pharmaceutical company, made public her findings that a particular drug was harmful. This violated the terms of her contract, and so she was dismissed. This caused a scandal, and she was subsequently reinstated. The university and hospital in question are now working out something akin to tenure for hospital-based researchers and new guidelines for contracts, so that more public disclosure of privately funded research will become possible. This is a rare (partial) victory and a small step in the right direction, but the general trend is the other way. Thanks to profit-driven private funding, researchers are not only forced to keep valuable information secret, they are often contractually obliged to keep discovered dangers to public health under wraps. Of course, we must not be too naive about this. Governments can unwisely insist on secrecy, too, as did the British Ministry of Agriculture, Fisheries, and Food in the work they funded in connection with the bovine spongiform encephalopathy (mad cow) epidemic. This prevented others from reviewing the relevant data and pointing out that problems were more serious than the government was letting on.

A recent study found that more than one-third of recently published articles produced by University of Massachusetts scientists had one or more authors who were likely to profit from the results they were reporting. They either were patent holders, or had some relationship to a company that would exploit the results, for example, as board members. The financial interests of these authors were not mentioned in the publications.[2] If patents are needed to protect public knowledge from private claims, then simply have the publicly funded patent holders put their patents in the public domain or charge no fee for use.

In another case, financial institutions donated a very large sum to a Canadian university economics department to study "the effects of high taxation on productivity." The results may influence government policy. In such cases, the public and its political decision-makers get information only of a certain kind, because there is no private, well-funded foundation called "The Consortium of Single Mothers on Welfare" that bestows similar massive funding to discover the effects of poverty on the development of children. Public policy decisions should be based on a variety of sources of information, but the privatization of research means that one point of view–guess whose?–will tend to prevail. Publicly funded science, though far from perfectly serving all interests, has at least a chance of serving more of them.

To raise funds, many universities have instituted a system of matching grants. If an endowed chair costs 2 million dollars to fund, a donor perhaps need only give 1 million, and the university will provide the rest. But where do these matching funds come from? Usually every university department loses a bit of its budget in order to build up a pool. Do they get it back in the form of an endowed chair? Some do and some do not. The applied fields and the headline-grabbing fields do well on this scheme, but the so-called pure sciences and especially the humanities are being decimated. A matching-funds scheme takes decision-making out of the hands of academics and gives it to donors. We may think that our limited resources should go to, say, Byzantine history or evolutionary biology, but applied research is more likely to be popular with donors who are now empowered by the matching grants procedure to redirect our limited funds.

When Derek Bok was president of Harvard University, he warned that strong leadership would be needed to protect our research goals from the eroding effects of commercial concerns. He was right to sound the alarm, but it will take a great deal more than strong leadership in the university. It will require massive government protection and promotion of public knowledge. Patent laws, for instance, must not allow the privatization of the public good. University research must be funded overwhelmingly from the public purse. And the public–rather than corporations or individual scientists (or even secretive governments)–must own the results.

To achieve this, regular academics must take up the cudgels. If they make an organized and concerted effort, academics could bring the current trend to a crashing halt. What can scientists do? At the individual level, they can refuse to do contract research that requires nondisclosure and insist on keeping knowledge public. At the university level, they can put pressure on administrators (who will sometimes welcome the support, since they, too, are deeply concerned) to take decision-making power out of the hands of private interests, corporate or philanthropic. At the political level, we can all pressure government leaders to keep research and education as part of the public good.

It is easy to fall into ideological debate on this issue, with one side upholding public knowledge for the sake of social justice and the other insisting on the value of private initiative and the need to financially reward it. However, there is a better way to view this cluster of issues; namely, in terms of efficiency.[3] The United States is unique among industrialized countries in not having a national health system. Health care is overwhelmingly private and largely in the hands of insurance companies. The cost is approaching 15 percent of the U.S. gross domestic product, and more than one-quarter of the population is not covered. By contrast, Canada (like most other industrialized countries) has universal coverage at a cost of under 9 percent of gross domestic product. Aside from the cost, it is hard to compare the relative quality of the health-care systems, but one statistic is revealing: Cancer patients in Canada live an average of 14 months longer from the time of detection than those in the United States.

The superiority of public health care is manifestly obvious; it is vastly more efficient, at least when properly funded, which is not always the case. Although there are disanalogies with research and education, a public health-care system can nevertheless serve as a model for how best to proceed. Why pay royalties to pharmaceutical companies when public research is more efficient? It's cheaper, safer, and better in every way.

Profit-driven medical research in the United States is (in many respects) topnotch, but is it the huge profits that make it so? Pure mathematical research in the United States is also topnotch, but publicly funded. No one could make a penny from Wiles's proof of Fermat's last theorem. Scientists need good salaries and the necessary resources, and they need to have their efforts appreciated. That is more than enough motivation for brilliant, effective science.

I do not for a moment believe scientists should be living in an ivory tower, indifferent to the world outside. The question is to whom they should be accountable–to use a favorite term of privatizers. The answer is simple: the public. Scientists owe it to them to keep knowledge free for all.

Science is the single most important institution in our lives. That claim ought to make us sit up and take notice–but it doesn't. We've become complacent. Coming to understand how science works and how it can be made to serve us better is surely–along with the elimination of poverty, to which it is connected–the first task for us all.

Notes

Bibliography

Index

Notes

1 Scenes from the Science Wars

1 Most recently, *The Sokal Hoax* was published. It contains Sokal's hoax paper and numerous pieces from the early days of the hoax. See Editors in the Bibliography.

2 Others such as Noretta Koertge wholeheartedly endorsed Sokal's actions. See Koertge's introduction (1998) for a statement of her views.

3 The term stems from Butterfield (1959). Whig historians were guilty, said Butterfield, of seeing British political history in terms of progressive steps toward current political liberty, rather than seeing those events in their own context.

4 I was editor of the journal *International Studies in the Philosophy of Science,* and also of a book series *Toronto Studies in Philosophy* (University of Toronto Press).

5 Gross and Levitt note that there are many on the Left who are also pro-science and people on the Right who are anti-science. Consequently, they recognize that the term "academic Left" is problematic for this reason. But their attempt to ameliorate the situation fails and the tone of their book is one in which we have the simple dichotomy: the anti-science Left stands opposed to the pro-science Right.

2 The Scientific Experience

1 To those not familiar with contemporary ethics, this may seem a farfetched example. It isn't. Much of current moral and political philosophy is based on the idea of negotiating a contract, and implicitly assumes that each party to the negotiations can best look after its own interests.

2 I follow Giere (1988 and 1991) in both historical detail and general moral concerning novel predictions.

3 See Giere (1991) for detailed discussions, both historical and philosophical.

3 How We Got to Where We Are

1 It can be found in Neurath (1973). The document was written in 1929 by Neurath, Carnap, and Hahn; other members of the Circle had some input, and they all endorsed it.
2 This is actually quite controversial. It's arguably true when we use a contemporary version of Newtonian mechanics and Keplarian laws. Historically, the issue is more subtle and the "derivation" is not valid.
3 See Howson and Urbach (1989) and Earman (1992) for further details of the Bayesian approach and its problems.

4 The Nihilist Wing of Social Constructivism

1 But there is room for rival interpretations of it. Most realists take the uncertainty to reflect our (necessary) ignorance, whereas anti-realists claim that the principle is about reality, so that if a momentum measurement is made, then there is no position. For this reason some anti-realists tend to prefer the term "indeterminacy principle," whereas realists prefer "uncertainty principle."
2 See Peirce (1992/98) or Misak (1995) for a history of verificationism.
3 Thanks to Bill Seager for this example.
4 Dadaism was an anti-art movement early in the twentieth century.

5 Three Key Terms

1 For a defense of Platonic realism in mathematics, see Brown (1999), and for realism in general, see Brown (1994).

6 The Naturalist Wing of Social Constructivism

1 See, for example, Jammer (1966).
2 For a look at one of the critics of Forman, see Hendry (1980).
3 So-called because Bloor and others such as Barry Barnes, Donald MacKenzie, and Steven Shapin were (at the time) all at the University of Edinburgh.
4 Merton (1968, 516). The principle is also embraced and discussed at length by Laudan (1977, 202).
5 Though Bloor worried little about reflexivity, others have made it almost a fetish. Those with a taste for it might consider Woolgar (1988). It's a tricky

business, however, and we will find ourselves coming back to reflexivity and related points repeatedly.

6 I cite this example, like the Forman example, to illustrate a way of understanding science. I won't spend any time going through the case and offering an alternative account. The case is discussed in further detail in Geison's recent biography of Pasteur (Geison 1995), and more briefly in Collins and Pinch (1993). Those interested in a critique of Farley and Geison should see Roll-Hansen (1979).

7 I'm making this overly simple, but the logical point should be clear, regardless.

8 For a full discussion with further references, see Gilbert (1989).

7 The Role of Reason

1 See, for example, the various papers in Kornblith (1985).

2 See, for example, Anscombe (1959), Dray (1957), Hampshire (1959), Meldon (1961), and Ryle (1949).

3 See, for example, Davidson (1963), and Kim (1995).

4 See, for instance, Zeller (1931) and Vlasotos (1971) for a debate on the matter over Plato.

5 The first version of this argument is in Nicholas (1984). I gave a different version in Brown (1989) to which Bloor replied in the Afterword to the second edition of *Knowledge and Social Imagery* (1991). The current exposition of the argument will, I hope, make it clear that his reply does not come to grips with the problem.

6 See, for example, Fleck et al. (1992).

7 See Brown (1991) for a Platonic account; Norton (1991) offers an empiricist version; Sorensen (1992) contains many examples; Gendler (1998) is perceptive; yet others can be found in Massey (1991).

8 The Democratization of Science

1 Both talks are reprinted in *The Sokal Hoax*. (See *Lingua Franca* (eds.) (2000) in the Bibliography.)

2 For preliminary details, see the article in the *New York Times* by Johnson (1996). This is the article that provoked Sokal in his exchange with Ross. More can be found in Anyon et al. (1996).

3 As reported in *The New Yorker,* Nov. 11, 1996.

4 For a brief history of the group, see Greenley and Tafler (1980).

5 This point was first made clearly in Okruhlik (1994).

9 Science with a Social Agenda

1 Quoted in Larson (1997, 44). My portrait of Bryan owes much to this interesting book.
2 See Gould (1999). The sympathy is for Bryan's hatred of Social Darwinism, but it does not, of course, extend to Bryan's views on evolution.
3 See Gould (1994) or the contributions to Fraser (1995) or to Devlin et al. (1997) for more criticisms of *The Bell Curve* and IQ studies in general.
4 We should reflect for a moment on what is taught to our children in the way of national history. I dare say the vast majority would say that history should be taught so that our children come to admire and appreciate their homeland. Yet such a view seems blind to the possibility that the historical truth might make our children ashamed. Most countries insist on teaching versions of events that are flattering. Germany and Japan, however, are under constant pressure to give their children a very uncomplimentary account. Other nations cheerfully pass over their own unsavory episodes. Perhaps they acknowledge that wrongs were done, but claim these were minor deviations from a pattern of glory. Bryan thought schools were for making citizens. If biology has to be fudged to do this, so be it. His view wasn't all that different from common views today about teaching history. For a sample of controversies in the case of the United States, see Harwit (1996) and Nash et al. (1998).

Afterword

1 The following is adapted from an opinion piece published in *Science,* "Privatizing the University–The New Tragedy of the Commons" (December 1, 2000).
2 For this and more such cases see Schulman (1999).
3 For more on the theme of efficiency in connection with public policy, see Heath (2001).

Bibliography

Anscombe, G. E. M. (1959) *Intention,* Oxford: Blackwell.

Anyon, R., T. J. Ferguson, L. Jackson, and L. Lane (1996) "Native American Oral Traditions and Archaeology," *SAA Bulletin* 14(2) (March/April), 14–16.

Arditti, R., P. Brennan, and S. Cavrak (eds.) (1980) *Science and Liberation,* Montreal: Black Rose Books.

Aronowitz, S. (1988) *Science as Power: Discourse and Ideology in Modern Society,* Minneapolis: University of Minnesota Press.

—— (1996) "The Politics of the Science Wars," *Social Text* 46–47(Spring-Summer), 177–196.

Baily, J. and R. Pillard (1991a) "A Genetic Study of Male Sexual Orientations," *Archives Gen. Psychiatry* 48, 1089ff.

—— (1991b) "Are Some People Born Gay?" *New York Times* (Dec. 17), A21.

Barnes, B. (1982) *T. S. Kuhn and Social Science,* New York: Columbia University Press.

Barnes, B. and D. Bloor (1982) "Relativism, Rationalism, and the Sociology of Knowledge," in S. Lukes and M. Hollis (eds.), *Rationality and Relativism,* Oxford: Oxford University Press.

Barry, A. (1898) *A Short History of Astronomy* (rpt.), New York: Dover.

Barsky, R. (1999) "Intellectuals on the Couch: The Sokal Hoax and Other *Impostures intellectuelles,*" *Substance* 88, 105–120.

Bauman, Z. (1989) *Modernity and the Holocaust,* Ithaca, NY: Cornell University Press.

Beller, M. (1998) "The Sokal Hoax: At Whom Are We Laughing?" *Physics Today* 51(9, Sept.).

Bernstein, J. (1996) *A Theory for Everything,* New York: Springer-Verlag.

Bloor, D. (1976/91) *Knowledge and Social Imagery* (2nd ed.), Chicago: University of Chicago Press.

Bloor, D., B. Barnes, and J. Henry (1996) *Scientific Knowledge: A Sociological Analysis,* Chicago: University of Chicago Press.

Boghossian, P. (1996) "What the Sokal Hoax Ought to Teach Us," *Times Literary Supplement* (Dec. 13), 14–15.

Boyer, C. (1959) *The Rainbow* (rpt. 1987), Princeton: Princeton University Press.

Bricmont, J. and A. Sokal (1997a) *Impostures Intellectuelles,* Paris: Editions Odile Jacob.

—— (1997b) "What is All the Fuss About?" *Times Literary Supplement* 4933 (Oct. 17), 17.

—— (1998) *Fashionable Nonsense: Postmodern Intellectuals' Abuse of Science* (trans. of 1997a), New York: Picador.

Brown, J. R. (1989) *The Rational and the Social,* London and New York: Routledge.

—— (1991) *Laboratory of the Mind: Thought Experiments in the Natural Sciences,* London and New York: Routledge.

—— (1994) *Smoke and Mirrors: How Science Reflects Reality,* London and New York: Routledge.

—— (1999) *Philosophy of Mathematics: An Introduction to the World of Proofs and Pictures,* London and New York: Routledge.

—— (2000) "Privatizing the University–The New Tragedy of the Commons," *Science* (Dec. 1, 2000).

Brown, J. R. (ed.) (1984) *Scientific Rationality: The Sociological Turn,* Dordrecht: Reidel.

Bunge, M. (1996) "In Praise of Intolerance to Charlatanism in Academia," in Gross, Levitt, and Lewis (1996).

Butterfield, H. (1959) *The Whig Interpretation of History,* London: Penguin.

Callon, M. (1986) "Some Elements of a Sociology of Translation: Domestication of the Scallops and the Fishermen of St. Brieuc Bay," in J. Law (ed.), *Power, Action, and Belief,* London: Routledge, 196–233.

Chomsky, N. "Rationality/Science" (from *Z Papers* Special Issue), at http://www.zmag.org/zmag/articles/chompomoart.html [accessed April 2001].

Cohen, I. B. (1985) *The Birth of a New Physics,* New York: W.W. Norton.

Collins, H. (1985) *Changing Order: Replication and Induction in Scientific Practice,* London: Sage.

Collins, H. and S. Yearley (1992) "Epistemological Chicken," in Pickering (ed.) (1992).

Collins, H. and T. Pinch (1993) *The Golem: What Everyone Should Know about Science,* Cambridge: Cambridge University Press.

Curd, M. and J. Cover (eds.) (1998) *Philosophy of Science: The Central Issues,* New York: Norton.

Davidson, D. (1963) "Actions, Reasons, and Causes," reprinted in *Actions and Events,* Oxford: Oxford University Press, 1981.

Dean, J. (1998) *Aliens in America: Conspiracy Cultures from Outerspace to Cyberspace,* Ithaca, NY: Cornell University Press.

Devlin, B., S. Fienburg, D. Resnick, and K. Roeder (eds.) (1997) *Intelligence, Genes, and Success: Scientists Respond to* The Bell Curve, New York: Springer-Verlag.

Dray, W. (1957) *Laws and Explanation in History,* Oxford: Oxford University Press.

Earman, J. (1992) *Bayes or Bust,* Cambridge, MA: MIT Press.

Eichler, M. and J. Lapoint (1985) *On the Treatment of the Sexes in Research,* Ottawa: Social Sciences and Humanities Research Council of Canada.

Einstein, A. (1944) "Russell's Theory of Knowledge," in P. A. Schilpp (ed.), *The Philosophy of Bertrand Russell,* New York: Harper and Row.

Farley, J. and G. Geison (1974) "Science, Politics, and Spontaneous Generation in 19th Century France," *Bulletin of the History of Medicine* 48, 161–198.

Feyerabend, P. (1962) "Explanation, Reduction, and Empiricism," in H. Feigl and G. Maxwell (eds.), *Minnesota Studies in the Philosophy of Science,* Minneapolis: University of Minnesota Press.

—— (1975) *Against Method,* London: New Left Books.

—— (1978) *Science in a Free Society,* London: New Left Books.

—— (1987) *Farewell to Reason,* London: Verso.

Feynman, R. (1985) *QED: The Strange Theory of Light and Matter,* Princeton: Princeton University Press.

Fish, S. (1996) "Professor Sokal's Bad Joke," *New York Times* (May 21).

Fleck, K., N. Cartwright, T. Uebel, and J. Cat (1992) *Otto Neurath: Between Science and Philosophy,* Cambridge: Cambridge University Press.

Forman, P. (1971) "Weimar Culture, Causality and Quantum Theory, 1918–1927: Adaptation by German Physicists and Mathematicians to a Hostile Intellectual Environment," *Historical Studies in the Physical Sciences* 3, 1–115.

Fraser, S. (ed.) (1995) *The Bell Curve Wars,* New York: Basic Books.

Friedman, M. (1998) "On the Sociology of Scientific Knowledge and Its Philosophical Agenda," *Studies in History and Philosophy of Science* 29, 239–271.

Fuller, S. (1999) *The Governance of Science,* Milton Keynes: Open University Press.

Gardner, M. (1996) "Physicist Alan Sokal's Hilarious Hoax," *Skeptical Inquirer* (Nov./Dec.), 14–16.

Geison, G. (1995) *The Private Science of Louis Pasteur,* Princeton: Princeton University Press.

Gendler, T. (1998) "Galileo and the Indispensability of Scientific Thought Experiment," *British Journal for the Philosophy of Science* 49(3), 397–425.

Giere, R. (1988) *Explaining Science,* Chicago: University of Chicago Press.

—— (1991) *Understanding Scientific Reasoning* (3rd ed.) Chicago: Holt, Rinehart, and Winston.

—— (1996) "The Feminism Question in the Philosophy of Science," in L. H. Nelson and J. Nelson (eds.) (1996).

Gilbert, M. (1989) *On Social Facts,* Princeton: Princeton University Press.

Gillispie, C. (1960) *The Edge of Objectivity,* Princeton: Princeton University Press.

Gould, S. J. (1994) "Curveball," reprinted in Fraser (ed.) (1995).

—— (1999) *Rocks of Ages: Science and Religion in the Fullness of Life,* New York: Ballantine.

Graham, L. (1985) "The Socio-political Roots of Boris Hessen: Soviet Marxism and the History of Science," *Social Studies of Science.*

Greenley, K. and S. Tafler (1980) "History of Science for the People: A Ten Year Perspective," in R. Arditti, P. Brennan, and S. Cavrak (eds.) (1980).

Gross, P. and N. Levitt (1994) *The Higher Superstition: The Academic Left and Its Quarrels with Science,* Baltimore: Johns Hopkins University Press.

Gross, P., N. Levitt, and M. Lewis (eds.) (1996) *The Flight from Science and Reason,* New York: New York Academy of Sciences.

Gutting, G. (ed.) (1980) *Paradigms and Revolutions,* Notre Dame: University of Notre Dame Press.

Hacking, I. (1999) *The Social Construction of What?* Cambridge, MA: Harvard University Press.

Halley, J. (1994) "Sexual Orientation and the Politics of Biology: A Critique of the Argument from Immutability," *Stanford Law Review,* 46–3, 503–568.

Hampshire, S. (1959) *Thought and Action,* London: Chatto and Windus.

Harding, S. (1986) *The Science Question in Feminism,* Ithaca, NY: Cornell University Press.

Harding, S. (ed.) (1993) *The "Racial" Economy of Science,* Bloomington: Indiana University Press.

Harwit, M. (1996) *An Exhibit Denied: Lobbying the History of* Enola Gay, New York: Springer-Verlag.

Hayles, K. (1992) "Gender Encoding in Fluid Mechanics: Masculine Channels and Feminine Flows," *Differences* 4(2), 16–44.

Heath, J. (2001) *The Efficient Society,* Toronto: Penguin.

Hendry, J. (1980) "Weimar Culture and Causality," *History of Science* 18.

Henley, J. (1997) "Is Modern French Philosophy Just a Load of Pseudo-Scientific Claptrap?" *The Guardian Weekly* (Oct. 12).

Herrnstein, R. J. and C. Murray (1994) *The Bell Curve: Intelligence and Class Structure in American Life,* New York: Free Press.

Hessen, B. (1931/71) *The Social and Economic Roots of Newton's* Principia, New York: H. Fertig.

Horwich, P. (ed.) (1993) *World Changes,* Cambridge, MA: MIT Press.

Howson, C. and P. Urbach (1989) *Scientific Reasoning: A Bayesian Approach,* La Salle, IL: Open Court.

Jammer, M. (1966) *The Conceptual Development of Quantum Mechanics,* New York: McGraw-Hill.

Johnson, G. (1996) "Indian Tribes' Creationists Thwart Archeologists," *New York Times* (Oct. 22).

Kamiya, G. (1996) "Transformative Hermeneutics of Total Bullshit," *Salon* (May 17).

Keller, E. F. (1996) "Science and Its Critics," in L. Menand (ed.), *The Future of Academic Freedom,* Chicago: Chicago University Press.

Kim, J. (1995) *The Philosophy of Mind,* Boulder: Westview.

Kimball, R. (1996) "'Diversity,' 'Cultural Studies' & Other Mistakes," *The New Criterion* (May).

Koertge, N. (ed.) (1998) *A House Built on Sand: Exposing Postmodernist Myths about Science,* Oxford: Oxford University Press.

Kornblith, H. (ed.) (1985) *Naturalizing Epistemology,* Cambridge, MA: MIT Press.

Krimsky, S. and L. S. Rothenberg (1998) "Financial Interest and Its Disclosure in Scientific Publications," *Journal of the American Medical Association* 280, 225–226.

Kuhn, T. S. (1957) *The Copernican Revolution,* Chicago: University of Chicago Press.

—— (1962/70) *The Structure of Scientific Revolutions,* Chicago: University of Chicago Press.

—— (1977a) *The Essential Tension,* Chicago: University of Chicago Press.

—— (1977b) "Objectivity, Value Judgement, and Theory Choice," in Kuhn (1977a), 320–339.

—— (1978/86) *Black-Body Theory and the Quantum Discontinuity, 1894–1912,* Chicago: University of Chicago Press.

Kusch, M. (1995) *Psychologism: A Case Study in the Sociology of Philosophical Knowledge,* London: Routledge.

—— (1999) *Psychological Knowledge: A Social History and Philosophy,* London: Routledge.

Lakatos, I. (1970) "Falsification and the Methodology of Scientific Research Programmes," in I. Lakatos and A. Musgrave (eds.), *Criticism and the Growth of Knowledge,* Cambridge: Cambridge University Press.

Larson, E. J. (1997) *Summer for the Gods: The Scopes Trial and America's Continuing Debate over Science and Religion,* Cambridge, MA: Harvard University Press.

Latour, B. (1987) *Science in Action: How to Follow Scientists and Engineers through Society,* Cambridge, MA: Harvard University Press.

—— (1988) *The Pasteurization of France* (trans. A. Sheridan and J. Law from *Les Microbes: Guerre et Paix suivi de Irréductions,* Paris, 1984), Cambridge, MA: Harvard University Press.

Latour, B. and S. Woolgar (1979) *Laboratory Life: The Social Construction of Scientific Facts,* London: Sage.

Laudan, L. (1977) *Progress and Its Problems,* Berkeley: University of California Press.

—— (1981) *Science and Hypothesis,* Dordrecht: Reidel.

—— (1984) *Science and Values,* Berkeley: University of California Press.

—— (1990) *Science and Relativism,* Chicago: University of Chicago Press.

—— (1996) *Beyond Positivism and Relativism,* Boulder, CO: Westview Press.

Leavis, F. R. and M. Yudkin (1962) *Two Cultures? The Significance of C. P. Snow,* London: Chatto and Windus.

LeVay, S. (1991) "A Difference in Hypothalamic Structure Between Heterosexual and Homosexual Men," *Science* 253, 1034ff.

Lewontin, R. (2000) *It Ain't Necessarily So,* New York: New York Review of Books.

Lingua Franca (eds.) (2000) *The Sokal Hoax: The Sham that Shook the Academy,* Lincoln, NB: University of Nebraska Books.

Lloyd, G. (1984) *The Man of Reason: "Male" and "Female" in Western Philosophy,* Minneapolis: University of Minnesota Press.

Longino, H. (1990) *Science as Social Knowledge,* Princeton: Princeton University Press.

Lyotard, J-F. (1983) "What Is Postmodernism," reprinted in T. Docherty (ed.), *Postmodernism: A Reader,* New York: Columbia University Press.

Mach, E. (1960) *The Science of Mechanics,* La Salle, IL: Open Court.

Martin, E. (1991) "The Egg and the Sperm: How Science has Constructed a Romance Based on Stereotypical Male-Female Roles," *Signs* 16.

—— (1992) *The Woman and the Body: A Cultural Analysis of Reproduction* (2nd ed.), Boston: Beacon Press.

Marx, K. (1859) *A Contribution to the Critique of Political Economy* (trans. from the German original, 1970), New York: International Publishers.

McMillen, L. (1997) "The Science Wars Flare at the Institute for Advanced Study," *The Chronicle of Higher Education* (May 16), A13.

Meldon, A. I. (1961) *Free Action,* London: Routledge and Kegan Paul.

Merton, R. K. (1968) *Social Theory and Social Structure,* New York: Free Press.

—— (1970) *Science, Technology and Society in Seventeenth-Century England* (originally published in 1938), New York: Harper & Row.

Misak, C. (1995) *Verificationism: History and Prospects,* London: Routledge.

Mulkay, M. (1979) *Science and the Sociology of Knowledge,* London: Allen & Unwin.

Nash, G. B., C. Crabtree, and R. E. Dunn (1998) *History on Trial: Culture Wars and the Teaching of the Past,* New York: Alfred A. Knopf.

Nelson, L. H., and J. Nelson (eds.) (1996) *Feminism, Science, and Philosophy of Science,* Dordrecht: Kluwer.

Neurath, O. (1973) *Empiricism and Sociology,* M. Neurath and R. S. Cohen (eds.), Dordrecht: Reidel.

Nicholas, J. (1984) "Scientific and Other Interests," in J. Brown (ed.), *Scientific Rationality: The Sociological Turn,* Dordrecht: Reidel.

Okruhlik, K. (1994) "Gender and the Biological Sciences," *Canadian Journal of Philosophy,* (supp. vol. 20), 21–42 (rpt. in Curd and Cover 1998).

Peirce, C. S. (1992/98) *The Essential Peirce* (2 vols.) N. Houser and C. Kloesel (eds.), Bloomington: Indiana University Press.

Pickering, A. (1995) *The Mangle of Practice,* Chicago: University of Chicago Press.

Pickering, A. (ed.) (1992) *Science as Practice and Culture,* Chicago: University of Chicago Press.

Pollitt, K. (1996) "Pomolotive Cocktail," *The Nation* (June 10).

Roll-Hansen, N. (1979) "Experimental Method and Spontaneous Generation: The Controversy Between Pasteur and Pouchet, 1859–64," *Journal of the History of Medicine* 34: 273–292.

Rose, H. and R. Rose (1970) *Science and Society,* New York: Penguin.

Rosen, R. (1996) "A Physics Prof. Drops a Bomb on the Faux Left," *Los Angeles Times* (May 23).

Ross, A. (1996) "Introduction," *Social Text* 46–47(Spring/Summer), 1–13.

Ross, A. (ed.) (1997) *Science Wars,* Durham, NC: Duke University Press.

Ruse, M. (1999) *Mystery of Mysteries: Is Evolution a Social Construction?* Cambridge, MA: Harvard University Press.

Russell, B. (1940) *An Inquiry into Meaning and Truth,* London: Allen and Unwin.

Ryle, G. (1949) *The Concept of Mind,* New York: Barnes and Noble.

Sagan, C. (1996) *A Demon-Haunted World,* New York: Ballantine Books.

Schatten, G. and H. Schatten (1983) "The Energetic Egg," *Sciences* (Sept./Oct.).

Schiebinger, L. (1999) *Has Feminism Changed Science?* Cambridge, MA: Harvard University Press.

Schulman, S. (1999) *Owning the Future,* New York: Houghton Mifflin.

Scott, J. (1996) "Postmodern Gravity Deconstructed, Slyly," *New York Times* (May 18).

Searle, J. (1995) *The Construction of Social Reality,* New York: The Free Press.

Shapere, D. (1965) "Meaning and Scientific Change," in R. G. Colodney (ed.), *Mind and Cosmos,* Pittsburgh: University of Pittsburgh Press.

Shapin, S. (1975) "Phrenological Knowledge and the Social Structure of Early Nineteenth-century Edinburgh," *Annals of Science,* xxxii.

—— (1996) *The Scientific Revolution,* Chicago: University of Chicago Press.

Shapin, S. and S. Schaffer (1985) *Leviathan and the Air Pump,* Princeton: Princeton University Press.

Slocum, S. (1975) "Woman the Gatherer: Male Bias in Anthropology," in R. Reiter (ed.) *Toward an Anthropology of Women,* New York: Monthly Review Press.

Snow, C. P. (1959/63) *The Two Cultures and the Scientific Revolution* (expanded version), Cambridge: Cambridge University Press.

Sokal, A. (1996a) "Transgressing the Boundaries: Toward a Transformative Hermeneutics of Quantum Gravity," *Social Text* 46–47, 217–252.

—— (1996b) "A Physicist Experiments with Cultural Studies," *Lingua Franca* (May-June), 62–64.

—— (1996c) "Transgressing the Boundaries: An Afterword," *Dissent* 43 (4), 93–99.

—— (1998) "What the *Social Text* Affair Does and Does Not Prove," in Koertge (1998).

Sokal, A. and J. Bricmont (1997) *Impostures Intellectuelles,* Paris: Editions Odile Jacob.

—— (1998) *Fashionable Nonsense* (rev. trans. of Sokal and Bricmont 1997), New York: Picador.

Sorensen, R. (1992) *Thought Experiments,* Oxford and New York: Oxford University Press.

Spengler, O. (1918/32) *Decline of the West* (English trans. 1932), London: George Allen and Unwin.

Stanley, L. and S. Wise (1983) *Breaking Out: Feminist Consciousness and Feminist Research,* London: Routledge and Kegan Paul.

Sullivan, P. (1998) "An Engineer Dissects Two Case Studies: Hayles on Fluid Mechanics and MacKenzie on Statistics," in Koertge (1998).

Tanner, N. and A. Zihlman (1976) "Women in Evolution, Part I," *Signs* I (3), 585–608 (see Zihlman 1978 for Part II).

Vlastos, G. (1971) "Reasons and Causes in Plato's *Phaedo,*" in G. Vlastos (ed.), *Plato,* Garden City, NY: Doubleday.

Weinberg, S. (1992) *Dreams of a Final Theory,* New York: Vintage.

—— (1996) "Sokal's Hoax," *New York Review of Books* (Aug. 8), 11–15.

Wheeler, J. and W. Zurek (eds.) (1983) *Quantum Theory and Measurement*, Princeton: Princeton University Press.

Will, G. (1996) "Physics Professor's 'Hilarious hoax' Reveals Intellectual Poverty of Deconstructionism," *Washington Post* (May 30).

Willis, E. (1996) "My Sokaled Life," *Village Voice* (June 25).

Woolgar, S. (1988) *Science: The Very Idea*, London: Tavistock.

Wylie, A. (1996) "The Constitution of Archaeological Evidence: Gender Politics and Science," in P. Galison and D. J. Stump (eds.), *The Disunity of Science*, Stanford: Stanford University Press.

—— (1997) "The Engendering of Archaeology: Refiguring Feminist Science Studies," *Osiris* 12, 80–99.

Zeller, E. (1931) *Outlines of the History of Greek Philosophy*, New York: Humanities Press.

Zihlman, A. (1978) "Women in Evolution, Part II," *Signs* IV (1), 4–20.

—— (1981) "Women as Shapers of Human Adaptation," in F. Dahlberg (ed.), *Woman the Gatherer*, New Haven: Yale University Press.

Zimmerman, B. et al. (1980) "People's Science," in Arditti, Brennan, and Cavrak (1980), 299–319.

Internet Sources

http://www.zmag.org/chomsky/ [accessed April 2001]

http://www.physics.nyu.edu/faculty/sokal/ [accessed April 2001]

http://www.math.tohoku.ac.jp/~kuroki/Sokal/index.html [accessed April 2001]

http://www.drizzle.com/~jwalsh/sokal/ [accessed April 2001]

Index